BUBBLES, BOOMS, AND BUSTS

The Rise and Fall of Financial Assets

Donald Rapp

BUBBLES, BOOMS, AND BUSTS

The Rise and Fall of Financial Assets

Copernicus Books

An Imprint of Springer Science+Business Media

Dr. Donald Rapp
1445 Indiana Avenue
South Pasadena, CA, 91030
USA
drdrapp@earthlink.net

Springer Science+Business Media, LLC © 2009

Published in the United States by Copernicus Books,
an imprint of Springer Science+Business Media.

Copernicus Books
Springer Science+Business Media
233 Spring Street
New York, NY 10013
www.springer.com

Library of Congress Control Number: 2008938905

Manufactured in the United States of America.
Printed on acid-free paper.

ISBN 978-0-387-87629-0   e-ISBN 978-0-387-87630-6

Preface

O ne of the problems that has challenged us for as long as we can remember is: how to value assets? In response to that challenge, we have invented the "free market economy" in which the price of an asset is set by the give-and-take between buyer and seller, one seeking the lowest price, and the other seeking the highest possible price. The two types of assets of greatest consequence to most of us are real estate and corporate stocks. According to classical economics, "the price is right" because it is set by negotiation between a rational buyer and a rational seller as to the "worth" of the asset. Unfortunately, history shows that at frequent intervals, this system gets seriously out of whack and the pricing of assets goes haywire. Stock and real estate prices are driven to "irrational exuberance." Inevitably, the bubble bursts and there is great misery throughout the land. Then the cycle repeats itself – again and again.

What seems to happen is that some event, some expectation, or some development starts the asset price rise rolling. As asset prices rise, a vacuum is generated that sucks in more investors, hungry for quick profits. The momentum so generated attracts more investors. By now, most new investors ignore the original stimulus for the boom, and are only buying with the intent of selling at a profit to "a bigger fool" who is expected to come along soon. Greed descends upon the land like a ground fog.

We have seen this process repeat itself with minor variations as far back as we can remember,[1] whether in tulips in Holland in the 17th century, the South Seas bubble of the 18th century, the Florida land boom of the 1920s, the stock market

---

[1] Early booms and busts were discussed in: McKay, Charles (1841), *Extraordinary Popular Delusions and the Madness of Crowds*. Richard Bentley, London. Reprinted Farrar, Strauss Giroux: New York: 1932

boom and crash of the 1920s, the great bull market of 1982–1995, the Japanese boom of the 1980s, the savings and loan scandal of the 1980s, the dot.com boom of 1996–2000, and most recently, the sub-prime mortgage housing boom of 2002–2007.

To add to the confusion, the bubble atmosphere provides a playground for charlatans, schemers, and crooks within which to operate. The Republican Party has provided impetus to these corporate criminals by implementing "deregulation" and interpreting this as "no regulation." In such an environment, banks and investment companies are free to play with the public's money and be bailed out by the Government.

The first part of this book examines the nature, causes and evolution of bubbles, booms and busts in asset markets as phenomena of human greed and folly. In doing this, I have built upon the foundations laid down by John Kenneth Galbraith's various works and I have also utilized material from Kindleberger's work: "Manias, Panics and Crashes," as well as various other sources cited in my book. Understanding bubbles, booms and busts requires first and foremost examination of the human element (greed, extrapolation, expectation and herd behavior).

The process by which a boom is transformed into bubble and thence to a bust is explored in considerable detail. In many cases, there is a legitimate basis for expecting significant future growth (as with the expansion of automobiles and highways in the 1920s, or the introduction and expansion of the personal computer and the Internet in the 1990s). This leads to investment, which produces a boom. The boom expands into a bubble when the original basis for investing is gradually replaced by *momentum buying* when speculators invest only because the asset price is rising. As prices rise, more speculators are sucked into the vacuum. Eventually, when the rate of rise reaches epic proportions, the bubble pops.

The rationality of investors comes into question. So does the rationality of bankers, who also display these same tendencies to an irrational degree. There is evidence that bankers are among the stupidest of people. Recent events in 2008 show that just about every major bank, brokerage house and mortgage company has been rocked by multi-billion dollar losses in the sub-prime mortgage fiasco, and their stock values have plummeted.

In addition, we examine how Government policy (monetary policy, fiscal policy, tax structure) – or the perception by investors regarding the effects of Government policy – affects bubble formation and collapse. Bubbles require money. The money is supplied by banks, which in turn are enabled by loose government monetary policies. Government policies include manipulation of interest rates and tax laws. Over the past thirty years, Government policies have been skewed repeatedly to support bubbles in real estate and stocks. Low interest rates hurt savers, and low income taxes (particularly on upper bracket income, capital gains and dividends) promote speculation and bubble formation. Asset bubbles enrich those who own assets. Therefore, it is relevant to examine who owns the assets in America. We find that a relatively few at the top own most of the assets. Hence preservation and enlargement of assets via bubbles preferentially benefits the rich, and that is, and has been the policy of the US Government. This raises the question whether asset bubbles create wealth, or vice versa? While classical economics might suggest that asset bubbles should merely create inflation, not wealth, there is considerable evidence in recent decades, that wealth has been created merely by bidding up the prices of stocks and housing (on paper), thus defying the laws of classical economics. As a result, the rich get richer (relative to the poor and middle class) and the disparity between the top and the bottom expands with time. The major supporter, architect and protector of bubbles over the past several decades has been Arthur Greenspan who used Federal Reserve policies to combat fragility in bubbles in almost every instance whenever it appeared.

A great proportion of the apparent prosperity of our times is illusory. First of all, much of the prosperity is confined to the rich. Most of the prosperity is due to asset growth and since the rich own most of the assets, they have profited the most. By contrast, real wages (adjusted for inflation) have been relatively flat for some time. Modifications to the income tax structure by Republicans have exacerbated this disparity. In addition to asset growth, a huge expansion in debt: federal, state, municipal and personal, has created the illusion of wealth. Ronald Reagan's introduction of "spend and borrow," as a new theme for the Republican Party over the past two decades, competes with the Democrat's "tax and spend" philosophy. Vice-President Cheney voiced the Republican viewpoint: "Deficits don't matter." The combination of (1) asset bubbles, (2) expansion of debt, and (3) temporary control of inflation by purchasing cheap goods from China (while losing our manufacturing industries and blue-collar jobs) seems to have worked – but this shaky house of cards could easily collapse, and likely will.

The second part of this book examines a number of specific boom-euphoria-bust cycles during the last 100 years. Most of the emphasis is on American bubbles but a few overseas bubbles are also included.

The Florida land boom of the 1920s ushered in the era of boom-bust cycles in the 20th century, when a single piece of property might trade six times in a single day with each purchase heaping promissory note upon promissory note until the whole thing collapsed.

The stock market in the late 1920s was a bubble in which stock prices rose incredibly from 1924 to 1929, and the general atmosphere was that of a gigantic bubble driven by euphoric investors, with heavy margin buying and leverage introduced via investment trusts. However, a number of learned articles have claimed that most stocks were not overpriced in 1929. There are many explanations for why the stock market collapsed in October 1929, and all of these provide insights; nevertheless an all-inclusive explanation has yet to be found. Similarly, the explanations for the ensuing depression of the 1930s are diverse, but it is still not entirely clear why the depression was so profound and so lengthy.

The savings and loan scandal of the 1980s was partly a bubble and partly out-and-out fraud, encouraged, supported and abetted by policies of the Reagan administration that blindly believed that deregulation (interpreted as no regulation) would solve an inherent problem of S&Ls in which their revenues from mortgages would no longer cover their costs when interest rates on deposits escalated. The cost of bailing out failing S&Ls could have been contained if the Reagan administration had acted in a timely fashion; but it didn't, and unseemly speculators and criminals took over the S&L industry while Mr. Reagan kept his head in the sand. In the end, the taxpayers paid for the debacle after Mr. Reagan left office.

The dot.com mania of the late 1990s was based on a sound intuition that the Internet would have a profound positive effect on communications, business efficiency and information storage and retrieval. However, the boom very quickly turned into euphoria as new companies were created daily and bid up to incredibly high prices. The valuations (stock price × number of shares outstanding), given to minor Internet businesses with no earnings, often exceeded valuations of major companies like General Electric. It was inevitable that after the huge run-up in stock prices prior to 2000, the bubble would collapse in 2000; and collapse it did with a "thud."

Mr. Greenspan tried to rescue the collapsing stock market with a series of drastic rate cuts starting in 2002, and to some extent he was successful. But an unintended (at least presumably unintended) consequence of the rate cuts was the generation of a new huge bubble in residential housing prices from 2002 to 2007. This bubble was aided and abetted by the prevailing interpretation of deregulation of banks and home loan institutions as "no regulation" – allowing them to pursue speculative, risky, and in many cases just plain stupid policies regarding issuing mortgages without adequate down payments, and issuing gerrymandered loans to people who could not afford the payments, in the expectation that rising house prices would bail them out. This was further exacerbated by large financial institutions packaging large numbers of mortgages into investment vehicles that obscured the fragility of the underlying collateral.

When the housing bubble popped in late 2007, as it had to, it dragged down the stock market as the realization spread that most financial institutions had lost countless billions in inflated real estate securities. However, once again "Helicopter Ben" and the Fed came to the rescue dropping down money on the markets after every significant falter in the stock market. And with each money drop, the dollar weakened, the price of oil shot up, and the price of gold inflated.

Perhaps most wondrous of all is not the repeated boom-bubble-bust cycle that we see over and over again in asset investments; but rather it is the almost religious belief of investors who prostrate themselves before the Federal Reserve with its rate-settings, as if like a Colossus astride the economy, it can single-handedly steer the ship of state to safety.

Contents

Contents

Contents

Introduction: The Holland Tulip Mania of 1636–1637

*H*istory shows that there is a deep-rooted human urge to make a quick profit without working for it by trading in paper assets. One of the first documented boom-bubble-bust cycles was the "tulip craze" that took place in Holland in 1636–1637 when buying and selling tulips became a national mania that led otherwise rational people into mortgaging their worldly goods to invest in tulips.

Tulips originated in Asia and Turkey, where they were cultivated and propagated in Turkey almost a thousand years ago. They were introduced into Holland for the first time in 1563, where they were propagated and studied by a Dutch botanist from the 1570s to the 1590s. The culture of tulips and propagation from bulbs or seed is a slow process. By 1600, tulips were in some demand throughout Europe but supplies were limited. The colors of tulips began to change due to a virus and some magnificent tulips evolved. Tulips were valued by their color, and a hierarchy of tulips evolved with the most desirable ones bringing very high prices. A tulip called *"semper augustus"* was the most highly prized of all, and quickly became very valuable.[2]

Between 1600 and 1630, Dutch tulip growers propagated more tulips, and tulip sales became a thriving business. Tulips were taken out of the ground after the blooming season and dried and stored for the summer to preserve them prior to replanting in the fall. Most sales therefore took place in mid to late summer when the bulbs were accessible. With the passage of time, tulip prices rose significantly, but in an orderly fashion.

In this era, some Hollanders became wealthy through trade with distant lands, but the great majority of the Dutch were artisans or farmers who worked long hours for subsistence wages. It was tempting to these laboring people to try to earn some additional money by acquiring and propagating tulips themselves. Thus,

---

[2] "Tulipomania", Mike Dash, Three Rivers Press, 1999.

with the expansion of the tulip market, a number of amateurs began growing tulips for sale in the early 1630s.

Dash described two national propensities of the Dutch of that time: savings and gambling. The plague killed off a number of people during the 1630s, leaving a shortage of labor. Wages went up as a result, and artisans had some extra savings to gamble on the tulip trade. Tulip prices rose considerably from 1630 to 1635, and the interest in earning profits from tulips expanded amongst the populace during that period.

The demand for tulips was such that a market that only existed for about two months in late summer was inadequate. As a result, in 1635, an important change was made in the way tulip sales were carried out. Instead of an exchange of cash for bulbs in late summer, the transactions could now take place at any time of the year, even while the tulip bulbs remained in the ground, and the exchange of cash was for a contract in which the bulbs would be made available to the buyer at the next late summer opportunity. This introduced several issues because the buyer was not sure exactly what he was getting, and the care of the sold bulbs was not always ideal. At the same time, many sales were made on contracts in which the buyers put up little cash, but paid a down payment in kind, with personal goods, and promised to pay the seller a large cash payment after the buyer took possession (based on the expectation that he could sell the bulbs to another buyer at a higher price). Most of these people could not possibly come up with the cash required at completion of the deal, except by selling their tulips to a hypothetical future buyer.[3] Very often, the down payment was a small percentage of the total price. Thus, buyers were highly leveraged. With these changes in the market, there was no need to know much about growing or propagating tulips. Investments were now made for the purpose of resale, not for the purpose of use. Thus, the tulip market passed from a boom phase to a mania phase.

Beginning in the autumn of 1635, prices escalated and as they did, more and more investors were sucked into the market to buy, driving prices higher and higher. By 1636, tulips were traded on the stock exchanges of numerous Dutch towns and cities. This encouraged trading in tulips by all members of society, with many people selling or trading their other possessions in order to speculate in the tulip market. By the autumn of 1636, a single tulip bulb could command a price equivalent to a few years' average salary, and the top bulbs were priced at several decades of average salary. Prices rose by a factor of ten from November 1636 to

---

[3] If this sounds familiar in 2008, it is because this was the same philosophy of those who bought housing that they could not afford during 2004–2007 with the expectation that rising prices would bail them out.

January 1637. The peak in the market occurred in early February 1637, when an auction of tulips netted 90,000 Guilders.[4] However, at an auction a few days later, there were no bids. This led to a nationwide panic as buyers disappeared from the markets. The ensuing collapse of the tulip market was swift and profound. By the spring of 1637, tulip prices had dropped by factors of 20 to 100. Many of the relatively common tulips became completely worthless. As in the case of the Florida land boom of the 1920s, a given tulip may have been bought and sold several times, each time with a small down payment and a promissory note. As each buyer defaulted, they left behind a tangled web of unpaid bills.

Had the tulip transactions been enforced, those who had mortgaged their few possessions to enter the tulip market would have been ruined – implying consignment to the workhouse, or starvation. Attempts were made to resolve the situation to the satisfaction of all parties, but these were unsuccessful. Ultimately, individuals were stuck with the bulbs they held at the end of the crash—no court would enforce payment of a contract, since judges regarded the debts as contracted through gambling, and thus, not enforceable by law. In many cases the people who owed had no assets worth suing for anyway. It appears that after the collapse of the tulip market, the courts decreed that all purchase contracts would be treated as options to buy and need not be fulfilled.

Dash described the end result of the tulip craze as surprisingly benign. Most of the crazy deals were negated and life went on, although bankruptcies increased and there are other signs of financial stress in the aftermath. However, Galbraith claimed that a recession followed the puncture of the bubble.

---

[4] For calibration, an artisan's salary was about 300–400 Guilders/year and a prosperous merchant may have earned 1,000 or more Guilders per year.

Chapter 1

THE NATURE OF MANIAS, BUBBLES, AND CRASHES

INTRODUCTION

John Kenneth Galbraith (JKG)[1] pointed out:

> The free-enterprise economy is given to recurrent episodes of speculation. These great events and small, involving bank notes, securities, real estate, art, and other assets or objects are, over the years and centuries, part of history.

*H*e then sought to find common features for these episodes because as he said, only through such understanding can the investor be warned and saved from "what must conservatively be described as mass insanity." However, as JKG amply demonstrated, such warnings will be met with vilification and abuse by the ruling powers during the manic phase of a boom.

JKG concluded:

> The more obvious features of the speculative episode are manifestly clear [in which assets] when bought today, are worth more tomorrow. This increase and the prospect attract new buyers; the new buyers assure a further increase. Yet more are attracted; yet more buy; the increase continues. The speculation building on itself provides its own momentum.

JKG described two types of participants in these booms. The *true believers* "are persuaded that some new price-enhancing circumstance is in control, and they expect the market to stay up and go up, perhaps indefinitely." They envisage a brave new world ahead where the rules have changed. A smaller group of super-ficially more *astute speculators* are aware of the speculative mood of the moment and the likelihood that it will eventually come to an end. Their goal is to ride the

D. Rapp, *Bubbles, Booms, and Busts*, DOI 10.1007/978-0-387-87630-6_1,
© Springer Science+Business Media, LLC 2009

upward wave with the aim of getting out before the speculation runs its course. If they are successful, they will do very well.

It is in the nature of speculative booms that there will be an inevitable fall, and that fall will not usually occur gently or gradually. It will typically occur with a comparatively sudden and decisive collapse. According to JKG, "that is because of the groups of participants in the speculative situation are programmed for sudden efforts at escape." At some point in the upward spiral, JKG postulated that "something, it matters little what—although it will always be much debated—will trigger the ultimate reversal." Astute (or lucky) speculators may get out in time. Those who thought the wave would continue upward forever ride the downward wave until they sell in desperation, driving the market down further. And as JKG summarized:

> . . . thus the rule, supported by the experience of centuries: the speculative episode always ends not with a whimper but with a bang. There will be occasion to see the operation of this rule frequently repeated.

The mathematics of rise and fall may not be widely understood. When the price of an asset doubles (increases by 100%) it only requires a 50% drop to restore it to its original price. Thus a stock that was originally priced at 10 that doubles to 20 needs only to drop by 50% to return it to 10. Similarly, if an asset goes from 10 to 40, a 400% appreciation, then a mere 50% reduction will wipe out half of the 400% gain. Thus, even a moderate drop from the high point can erase a substantial percentage of the previous gains. Those who joined the boom late may be hit especially hard by such losses.

JKG commented on the mass psychology of the speculative mood. In order for an individual to resist the suction generated by the allure of quick riches during a speculative boom, he must:

> . . . resist two compelling forces: one, the powerful personal interest that develops in the euphoric belief, and the other, the pressure of public and seemingly superior financial opinion that is brought to bear on behalf of such belief.

In this connection, JKG quoted Schiller's dictum:

> **Anyone taken as an individual is tolerably sensible and reasonable but as a member of a crowd, he at once becomes a blockhead.**

Those involved with the speculation are experiencing an increase in wealth and there is a tendency for them to believe that this is neither fortuitous nor undeserved. "All wish to think that it is the result of their own superior insight or intuition."

According to JKG, two factors that contribute to the bubble mentality are:

(1) The short financial memory (or ignorance of history) that makes investors oblivious of previous financial disasters and

(2) The tendency to attribute greater intelligence to individuals, the more income or assets that they control.

The ignorance of the history of booms and busts is a theme that was oft repeated by JKG. He suggested that there are many characteristics in common between boom/bust cycles over the past 400 years and "the lessons of history are compelling—and even inescapable."

In a world where acquisition of riches is difficult for most people, admiration for those who have accumulated wealth is seemingly unbounded. The public's fascination for the great financial mind is only dampened by speculative collapse, which then leads to disillusionment—until the next speculative boom.

A third factor (not discussed by JKG) that contributes to the bubble mentality is the expectation that the Government and central banks will "bail out" speculators through active intervention with monetary and fiscal policies if and when the bubble pops. There is ample evidence of this in the United States. Recent examples of this are the Government and Federal Reserve reactions to the popping bubble from speculation in real estate and stocks during 2002–2007 in which they flooded the banks with low-interest funds, printed money to distribute to the public, and provided support to those who speculated and over-borrowed.[2]

JKG described the aftermath of the end of "the inevitable crash" as "a time of anger and recrimination and also of profoundly unsubtle introspection." The anger will be directed against those who were previously respected as the most perceptive financial gurus. Some of them will have "gone beyond the law, and their fall and, occasionally, their incarceration will now be viewed with righteous satisfaction."

It would be of great value to investors if there were a good method to sense the oncoming of the end of a bubble. However, this would not apply to those who believe the bubble will have no end. Unfortunately, there does not seem to be any reliable process for sensing that the end is near. However, there does seem to be some indication that high volatility with wild swings upward and downward may (at least sometimes) presage the end of a stock bubble.

JKG went on to say there will be investigations into previous financial practices that were highly praised in their heyday. Some of these practices were merely implausible; others were clearly illegal. And as JKG indicated:

There will be talk of regulation and reform. What will not be discussed is the speculation itself or the aberrant optimism that lay behind it. Nothing is more remarkable than this: in the aftermath of speculation, the reality will be all but ignored.

JKG suggested that there are two reasons for this. One is that there are too many people and institutions involved and he emphasized:

Whereas it is acceptable to attribute error, gullibility, and excess to a single individual or even to a particular corporation, it is not deemed fitting to attribute them to a whole community, and certainly not to the whole financial community. Widespread naiveté, even stupidity, is manifest; mention of this, however, runs drastically counter to the earlier-noted presumption that intelligence is intimately associated with money.

According to JKG, the second reason that the speculative mood and mania are exempted from blame is that there is an almost theological faith in the free enterprise market "so there is a need to find some cause for the crash, however far fetched, that is external to the market itself." JKG cited the investigations and probes after the 1987 stock market meltdown, none of which ever considered excessive speculation as the main contributing factor.

Finally, JKG closed with a discussion of what, if anything, should be done. He suggested that there probably is not a great deal that can be done. It is impractical to attempt to outlaw mass financial euphoria that seems to be imbedded in the human psyche.

THE HUMAN ELEMENT

It seems to be a fundamental (perhaps even genetic) trait of humans that people appreciate a windfall more than almost anything else. Many people are very proud of the profit they made from a rise in their investment, particularly if some unforeseen event (a buyout?) drove a stock price well above what they had originally expected. By contrast, fewer people seem to find satisfaction in the hard-earned bucks they made from their salaries by dint of their labor. We seem to value investment income over wages. Perhaps that is why wages are taxed at a much higher rate than capital gains.[3] The data on growth of real wages seem to be contradictory. One source indicates that real average wages in America have slowly edged up in the last twenty-five years as shown in Table 1.1.

Table 1.1 Average real wages in America (constant 2005 dollars)[4]

Year	Wages
1967	$25,500
1973	$29,700
1979	$29,900
1989	$32,700
1995	$33,700
2000	$37,900
2004	$37,400

Another source[5] presented data on average hourly wages of production and non-supervisory workers. If these data are adjusted for inflation using the consumer price index, the real gain in wages from 1964 to 2005 (41 years) was a mere 4.5%. However, use of other measures of inflation leads to larger gains in wages. Nevertheless, real wages have risen slowly since 1973 and many middle-class Americans had to depend heavily on asset growth (mainly stocks and real estate) as well as two earners per family to get by in the late 20th century.

Markets in common stocks, real estate and other assets have provided investors with media for seeking paper profits from capital appreciation for hundreds of years.

One may conceive of hypothetical criteria for determining the worth of an asset. For example to obtain an estimate of the *worth* of a residence as the replacement cost, one could estimate an average construction cost in the local area ($/square foot) and multiply this by the number of square feet in the residence, and add this to an estimated land value. At different times, buyers are willing to pay more (or less) than the estimated replacement cost of a residence. In fact, during the housing bubble in California from 2001 to 2007, the connection between sales price and replacement cost was not typically a consideration. The value of a share of common stock is even more subjective. On a theoretical basis, the *value* or *worth* of assets such as common stocks and real estate is almost always quite subjective. In actual practice, the *value* at any moment can be construed to be what someone else is willing to pay for it.

History shows that all asset markets fluctuate as buyers and sellers move into or out of the markets. In some cases, now and again, a strong trend (upward or downward) may be established. This may be due to a random occurrence, or more likely, to some important outside factor (or factors) that exerts an influence. Kindleberger and Aliber (K&A),[6] following Minsky, suggested that an upward boom can be initiated by "an exogenous outside shock to the macroeconomic

5

system . . . if sufficiently large and pervasive." They suggested: "the rapid expansion of automobile production and associated development of highways together with electrification of much of the country . . . provided such a shock in the 1920s." The development and expanded use of the Internet in the late 1990s provided a shock that also produced a common stock boom. K&A also described a form of shock they called a "displacement" in which an outside event, typically unforeseen, "changes horizons, expectations, anticipated profit opportunities, . . ." They mentioned changes in the price of oil or outbreak of war as examples of shocks that produce displacement. K&A also suggested that the aftermath of a bubble generated by such a boom is usually a crash.

In 1995, when JKG republished his classic work "The Great Crash," he described the boom-bust cycle as follows:

> There is a basic and recurrent process. It comes with rising prices, whether of stocks, real estate, works of art or anything else. This increase attracts attention and buyers, which produces the further effect of even higher prices. Expectations are thus justified by the very action that sends prices up. The process continues; optimism with its market effect is the order of the day. Prices go up even more. Then, for reasons that will endlessly be debated, comes the end. The descent is always more sudden than the increase; a balloon that has been punctured does not deflate in an orderly way.

He also emphasized: "at some point in the growth of a boom all aspects of property ownership become irrelevant except the prospect for an early rise in price." Any use of the enterprise, or its value in the long run becomes academic. Instead, the only concern becomes whether the market price will rise soon, as it has in the recent past. There is no other benefit to ownership than the hope of selling at a higher price in the near future. In fact, JKG suggested that if the actual business conducted by the enterprise could somehow be divorced from the "burdens of ownership, this would be much welcomed by the speculator. Such an arrangement would enable him to concentrate on speculation which, after all, is the business of a speculator."

While there are many theories, it is difficult to be certain how or why such a boom originates. The important point is that whether due to random fluctuations or exogenous shocks, moderate upward movements in the prices of assets occur rather frequently. In most cases, the natural laws of supply and demand dampen these movements, leading to limited oscillations about the long-term trend line as shown in Figure 1.1. In a few cases, a boom develops in which prices rise unaccountably, eventually reaching extraordinarily high values, as shown in Figure 1.2.

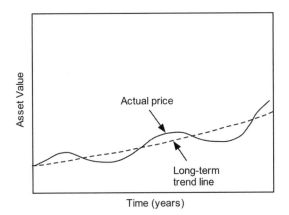

Figure 1.1 "Normal" fluctuations of asset value about a long-term trend line.

When such an upward trend begins, it provides a great attraction to many people who see this trend as a pathway to riches. While many hold back at first, as the boom accelerates upward, the urge to join in becomes almost irresistible.

A good metaphor is provided by the classic silent film of 1927, directed by René Clair: "A Nous La Liberté." There is a scene in the film in which many dignified members of the Chamber of Deputies, dressed in formal attire, are lined up in a courtyard to honor an entrepreneur who is upstairs preparing to abscond with a suitcase full of paper money. A windstorm comes up and blows the bills out the window where they drop down and swirl around at the feet of the Deputies. At first, no one makes a move. Then one, then two, then a few Deputies start picking up bills. Finally they are all floundering around on all fours retrieving errant bills.

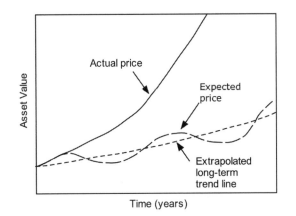

Figure 1.2 Market boom departs from a long-term trend line.

The essence of the boom is *momentum buying*. Whether it is stocks or real estate or whatever, the issue is no longer one of *value* in the usual sense. In this scheme of things, the whole notion of a *value* of an investment or commodity becomes irrelevant. Once such a trend is established, a large number of greedy investors jump on the bandwagon. By investing further into the boom, they create more demand, driving prices further up. As this upward momentum spiral expands, the lure of wealth, quick wealth, and wealth unearned becomes enormous. Those who held back in the beginning are sucked in, as by a gigantic vacuum.

As K&A pointed out astutely:

> "There is nothing as disturbing to one's well-being and judgment as to see a friend get rich."

Many more people cannot resist the temptation to make a quick buck for no effort, and more and more money pours into the system, driving prices to previously unimaginable levels. The urge to make quick unearned profits becomes so great that investors borrow to the hilt to invest even more funds into the boom, thus leveraging their investments.

Momentum buying involves: (1) identify a trend, (2) pay almost any price to get aboard the trend, (3) wait a bit for someone ("a bigger fool") to come along and drive up the price further, and (4) sell to him. Who has not received a circular urging purchase of a common stock at 50 that not so long ago was at 1 or 2?

Inevitably, the result of such a boom is that eventually, asset prices top out when they are driven to such extraordinary values that they can no longer be sustained. For example, housing prices may become so high that hardly anyone can afford to buy one, and the sales boom collapses. Or by some strange coincidence, a number of investors may feel that the boom has run its course, and sell, thus driving prices down. There may be more objective reasons for the end of the boom. For example, in the California housing boom of 2001–2007, many speculative house buyers took on adjustable rate mortgages with low initial rates, expecting that capital appreciation would allow them to sell out with a profit just about when the mortgage rate was programmed to increase to an unaffordable level. When capital appreciation topped out in 2007, they were left holding mortgages that stepped up to levels they could not afford. Some politicians rushed in to try to bail out these speculators under the belief that they were poor people manipulated by big bad banks.

Momentum selling works in reverse of *momentum buying*, although price drops tend to be more precipitous than price increases for a number of reasons (it requires funds to buy, not to sell; margin calls can produce forced selling of stocks; during downward spirals, there may not be buyers around, ...). The greatest challenge in *momentum buying* is step (4) selling to a bigger fool before the inevitable collapse of the bubble. The problem is that while a boom is racing upward, it is difficult to tell how far the market is from a top, and the euphoria is so endemic that few people perceive that there even will be a top. Very few people can sell out during the middle of a boom without great regret and heartache when prices continue to rise after they sold out. Many stories abound of investors who sold, agonized as the market continued upward, and then were lured back in, only to see the market collapse upon their second investment.

It is widely believed that loose monetary policy and low interest rates promote bubble formation. This is undoubtedly true to a great extent. However, JKG also concluded:

> Far more important than rate of interest and the supply of credit is the mood. Speculation on a large scale requires a pervasive sense of confidence and optimism and conviction that ordinary people were meant to be rich.

The effect of low interest rates does not seem to be a direct stimulation of business, as much as it discourages savers from investing in interest-bearing accounts and securities, thus promoting paper asset bubbles. JKG also reported on Will Payne's distinction between an investor and a gambler, saying that in gambling there is a fixed amount of money and there is a loser for each winner. However, in investing, if the bubble keeps expanding, everyone wins. A buys at 10, and sells at 20 to B. B sells to C at 30, etc. Until the bubble pops, everyone gains.

THE RISE OF MANIAS AND BUBBLES

It seems likely that manias and booms are not typically generated merely on the basis that an arbitrary market is rising due to a statistical fluctuation, and investors want to get "on board" before the train reaches its destination. In many cases, there seem to be external influences that lead investors to think that there is a basis in substance underlying the boom. These influences may be categorized as follows:

(1) *New technology*: K&A postulated (after Minsky) that "shocks" were responsible for booms. JKG suggested that discovery of something that is ostensibly new may provide the impetus for a new boom and bubble. These may be alternative ways of saying almost the same thing. Certainly, there have been booms and bubbles that were based on sound anticipation of gains to be made from new

technical developments. There may be beliefs that new technology will fundamentally alter the business picture, allowing companies to make unprecedented profits. This was a major factor in the 1920s boom (widespread expansion of automobiles, highways and electrification), and the 1990s (the belief that the Internet would change the way we purchase goods and communicate). In both instances, these beliefs proved to be correct in the longer run. Automobiles, highways, electrification, and the Internet have indeed eventually produced massive benefits to society and great increases in efficiency and productivity. Where investors went wrong was in anticipating a more immediate payoff than was possible, and more importantly, these fields became overpopulated with too many companies rushing in to participate too quickly. Ultimately, there had to be a shakeout that eliminated many of weakest. As the years went by, the stronger firms with the best products (e.g. Google) prospered. These booms were based initially on sound perceptions, even if the timing was off and the enthusiasm got out of hand. Nevertheless during the booms of the 1920s and 1990s, investors bid up the price of common shares to levels far beyond what could reasonably be considered appropriate (in retrospect), even taking into account the benefits of new technology.

(2) *Domino effect*: There is a phenomenon that K&A called "bubble contagion." According to K&A, four distinct asset price bubbles in the last fifteen years of the twentieth century were systematically related. As each bubble popped, the remaining accumulated funds found their way into an emerging bubble in another country. The Japanese real estate and stock bubble provided funds to expand the bubble in real estate and stocks of Finland, Norway, and Sweden. After the Japanese bubble burst around 1990, an inflow of funds from Tokyo in the several years following the Japanese implosion supported the bubbles in Thailand, Malaysia, and Indonesia and their neighbors in the mid 1990s. This eventually led to a surge in the flow of funds from the East Asian countries to the United States that helped boost the dot.com bubble of late 1990s in the United States. Asset price bubbles in major industrial countries had been rare prior to the last few decades of the 20th century, with the boom of the late 1920s representing a unique occurrence. However, beginning around 1982 we have experienced repeated bubbles, often with multiple nations participating.

(3) *Money supply and interest rates*: It is widely believed that actions (or inactions) of central banks via monetary policy and interest rate adjustment have a profound effect on the economy. For example, in late 2007 and 2008, whenever the US Federal Reserve System even hinted at a cut in interest rates, common stocks went shooting up, and in the absence of prospects for a rate cut, other financial woes weighed the stock markets down. The ensuing volatility

in the stock market was remarkable with many daily changes in the market averages of more than 2%.

The common belief is that the role of a central bank is a delicate balancing act. On the one hand, lowering interest rates and increasing the money supply improves prosperity, but it runs the risk of increased inflation. According to this theory, if it were not for the threat of inflation, central banks could reduce interest rates to generate almost any desired level of prosperity, but the threat of inflation inhibits this action. Typically, central banks lower interest rates when they fear stagnation in the economy. However, as this stimulus generates a boom, central banks are loath to stifle the boom by raising interest rates because of potential negative political consequences—they don't want to be the rain that falls on the investors' picnic. K&A discussed at some length whether monetary authorities should tighten credit to raise the cost of speculation during booms. They argued that when commodity and asset markets move together, up or down, the direction that monetary policy should take is clear (opposing extremes in either direction). But according to K&A, when share prices or real estate or both soar while commodity prices are stable or falling, the authorities face a dilemma. If they stifle speculation there is a likelihood that the economy could plunge. If they support the economy with low interest rates, speculation may be rampant. This dilemma was faced in the 1920s in the United States, in Japan in the late 1980s, and again in the United States in the late 1990s. Arthur Greenspan was concerned that United States stock prices were too high or increasing too rapidly when he made his famous remark "irrational exuberance" in December 1996. Despite this comment, the Federal Reserve was reluctant to raise interest rates to dampen the booming stock market because they were concerned that they might stifle economic growth. In addition, the Fed was concerned about the so-called "Y2K problem, the likelihood that US computer systems collapse because so many software programs were not designed to [accommodate] the transition to 2000." As a result, they pumped money into the banking system to promote liquidity in late 1999. As K&A said: ". . . the money had to go someplace so it fed stock market speculation."

Supposedly, central banks are most concerned with keeping inflation at 2% or lower, but should asset prices be included in the calculation of inflation rate? K&A pointed out: "in one view, asset prices should be incorporated into the general price level because, in a world of [supposedly] efficient markets, they hold a forecast of future prices and consumption. But this view assumes that asset prices are determined by the economic fundamentals and are not affected by herd behavior that leads to a bubble." K&A concluded that central bankers "have been exceedingly reluctant to attempt to deal with asset price bubbles or even to recognize that they exist or could have existed." The answer seems to be that they don't want to be the *Grinch who stole Christmas*. Central banks would rather allow a bubble to expand than be accused of opposing prosperity.

This discussion by K&A seems to neglect the fact that the Federal Reserve is a quasi-political body that keeps one eye on the upcoming election. People are happier during boom times and are more likely to reelect the current party in power. In late 2007, the stock market underwent a number of severe one-day precipitous drops in reaction to the collapse of the housing bubble and its effect on sub-prime mortgages and bank losses, but the Federal Reserve rose to each such occasion with rate cuts that produced equally sharp one-day reversals in the stock market indices. The Federal Reserve seemed to take on the role of the protector of stock market bubbles.

Whether the Federal Reserve should intercede to protect the profits of speculators is arguable. But both sides of the debate do not seem to doubt the power of monetary policy to affect economic growth.

However, Robert E. Lucas (Nobel laureate in economics) argued against the common belief that easy money policy with low interest rates boosts economic growth. Ever more empirical evidence suggests that monetary policy may be ineffective. For example, two decades of close to zero interest rates in Japan and Switzerland have been totally inadequate to provide any stimulus to their sluggish growth. According to Lucas, the only significant effect of increasing the money supply is increasing inflation, which slows down growth in the long run. So any attempt to boost growth through reducing interest rates is therefore ultimately counterproductive.

It seems likely that easy money policy with low interest rates does boost speculation, paper profits and bubbles, but that is not quite the same as prosperity—or is it? In "Wealth and Inflation," I discuss the point that classical economics would predict that flooding the marketplace with easy money should produce significant inflation as more money chases the same amount of goods. However, in the past few decades, particularly the 1990s and 2002–2007, we have seen great expansions of the money supply without severe inflation (as defined by the conventional Consumer Price Index). However, if the rise in asset values were added to consumer price inflation, that would change inflation indices dramatically. In addition, there were special circumstances holding a lid on the cost of consumer goods in the 1990s that may not be working so well in 2007–2008.[7]

(4) Developing new areas with favorable features: On occasion, the prospect of opening up new areas for living in favorable climates can provide the impetus for investment bubbles. For example, over the years, there have been land booms in the South Seas, Florida and California and the Southwestern United

States [see "Florida Land Boom of the 1920s," "The Bull Market of 1982–1995," "The Sub-Prime Real Estate Boom" in Chap. 2].

(5) *Financial Innovation*: JKG[8] claimed that some booms and bubbles have been based on financial (as opposed to technical) innovations. One example is the advent of holding companies (a.k.a. investment trusts) in the 1920s. The stockholders issued bonds and preferred stock, and used the proceeds to invest in other common stocks, but all this amounted to was increased leverage: a means of increasing the amount of money invested in the stock market compared to the investment made by common stockholders in the holding company. Investment trusts were described in some detail by JKG.[9] They were in some sense a precursor to modern mutual funds. These trusts provided ordinary citizens with a means of investing on a leveraged basis into a broad aggregate of common stocks that they would not have been able to afford if they had bought stocks directly.

Another example was the issuance of junk bonds in the 1980s with comparatively high interest rates for the purpose of raiding and taking over legitimate companies. A third example during the 1980s was the deregulation of S&Ls in the misguided belief that this could allow them to cope with the underlying problem of high current interest rates paid out on deposits vs. low rates of return on long-term mortgages. A fourth example was the deregulation of utilities that led to criminal manipulation of utility assets by the Enron Corporation and others in the 1990s.

However, JKG took a dim view of "financial innovation." He suggested that what the world celebrates as great financial innovations are actually small variations on past systems that have been forgotten due to the "short memory of financiers." As JKG described it:

> The world of finance hails the invention of the wheel over and over again, often in a slightly more unstable version. All financial innovation involves, in one form or another, the creation of debt secured to a greater or lesser adequacy by real assets.

Many other booms and bubbles were based on pure fluff, and amounted to little more than elaborate Ponzi schemes. These include John Law's Banque Royale and its Mississippi Company (1716–1720) to pursue putative gold deposits in the Louisiana Territory in which the sale of stock was not used to prospect for gold but to pay French Government debts. The English version of the Ponzi scheme was the South Sea Company that also blew up in 1720, leading to passage of the "Bubble Act" to constrain illegitimate promotions. See "The New World" in

Chap. 2. The current debt of the US Federal Government in 2008 is a Ponzi scheme because there seems to be no way that these loans can be repaid except by borrowing via new loans.

FUELING THE BOOM

K&A discussed a number of factors that help fuel booms. Books, magazines, news reports and the media in general amplify the enthusiasm for the boom. JKG noted that during the boom years of the 1920s, there were many articles and press releases encouraging the bubble. One example was an article in the Ladies Home Journal entitled: "Everybody Ought to be Rich." The book "Japan as Number One—Lessons for America" was a 1979 best seller that "launched a thousand other efforts in *Japan Hyping*." In the 1980s, there were many rooms in the United States filled with white collar professionals paying hundreds of dollars each, for the right to hear a speaker tell of the wonders of Japanese management schemes that were far ahead of the rest of the world. The World Bank published "The East Asian Miracle" several years before the bubble in real estate prices and stock prices in Thailand and Malaysia and their neighbors imploded. As the boom progresses, the urge to create landmark skyscrapers and other buildings provides further fuel for the boom.

During bubble expansions, anyone who suggests that the bubble will pop is typically denounced in the press. On September 5, 1929, Roger Babson made his famous prognostication:

> Sooner or later, a crash is coming, and it may be terrific . . .. Factories will shut down . . . men will be thrown out of work . . . the vicious circle will get into full swing and the result will be a serious business depression.

Babson was roundly vilified by the whole investment community.

K&A asserted that rising asset prices (typically residential housing and stocks) provide positive feedback loops to national income, which then cyclically, further increases asset prices. According to K&A, households typically have savings or wealth objectives. As their paper wealth increases from the surge in asset prices, households save less from earned income because their future is secured by their increases in asset values, and thus they spend more for consuming goods and services. When stock prices increase, firms can raise cash from existing and new investors at lower costs and can undertake new projects that would be less profitable.

The main question here is whether rising assets produce wealth or increased wealth produces rising asset prices. This is discussed in the next section.

WEALTH AND INFLATION

ASSET GROWTH VS. WEALTH: CAUSE AND EFFECT?

Imagine that we could all get together at once and decide that the price of all housing in America will double as of today. Since the major asset of many people in the middle class is their residence, this would almost double the net worth of many millions of people. Similarly, suppose we could increase the price of stocks by a factor of ten. That would benefit the rich the most, but it would also increase the net worth of many in the middle class, particularly those at the higher end. Now if costs of consumer goods and services did not change much as a result of these changes in asset prices, they could borrow against the increase in value of their homes, or sell some securities, and use the money so released to purchase many more goods and services than they could yesterday. Homeowners and stock investors would be substantially richer than they were yesterday.

Alternatively, suppose that the federal government prints a huge pile of cash and distributes to each citizen an amount of cash equal to his net worth.[10] Then each person would be twice as rich as he was previously.

According to Economics 101, that cannot (or at least it should not) happen. For example, the following quote is lifted from an Internet source:[11]

Inflationary money such as bankers create from thin air obviously does not increase wealth of a nation nor its real buying power, as their increase of the money supply is not accompanied by an increase of real goods or services. The nominal buying power such money provides to borrowers is merely diluted buying power. . . . Easy-money policy can never cause real growth, but merely creates a nominal illusion of progress. In the end real wealth can only be increased through increasing the availability of real goods and services, and the only way to increase production of tangible services and commodities is by working more or by producing more efficiently. And productivity can only be improved to a substantial extent through investment in better machines, superior techniques or improved infrastructure. So a policy aiming at real growth must therefore promote saving and investment, and certainly should not stimulate consumption. Easy money does the opposite: it promotes consumption, discourages saving, penalizes investment and productive contribution, in the long run all slowing down real growth; exactly the opposite of what it was set up to do.

This is the standard textbook explanation. To produce a "real" increase in wealth, one must "increase production of tangible services and commodities" rather than merely raising asset values or printing money. Merely increasing asset values or the money supply should theoretically cause more money to chase the same amount of goods, driving up consumer prices, causing inflation, with no real gain in wealth.

However, increases in asset values or the money supply do not occur instantaneously and it may require years for their impacts to take effect. Over a sustained period, it is conceivable that continued expansion of the money supply and/or growth of asset values might actually create demand which could stimulate increases in capacity as well as innovation that could lead to "increased production of tangible services and commodities." Nevertheless, the classical economic belief is that expansion of the money supply and/or growth of asset values do not *per se* increase wealth and prosperity.

Nevertheless, we have seen several asset bubbles apparently contribute to American wealth during the period 1980–2007 without a proportionate rise in consumer prices. For example, the Dow-Jones Industrial Average (DJIA) rose from 1,000 to 14,000 from 1985 to 2007, a factor of more than 13, while the Consumer Price Index (CPI) only doubled over that period. Similarly, the Case-Shiller Real Estate Index (CSREI) for 20 major metropolitan areas rose from 100 in 2000 to as high as 206 in 2006, and despite backing off in 2007 and 2008, the CSREI remains considerably higher than it was seven years ago. For some areas, such as Los Angeles, the peak in CSREI in 2006 was 274, indicating that on average, a house in Southern California appreciated by 174% from 2000 to 2007. We have seen no such rise in overall consumer prices.

Therefore, a case can be made that, contrary to economic theory, and indeed contrary to common sense, wealth can be created by increases in paper asset values. It seems to work. If we can only convince ourselves to create an asset bubble, it appears that consumer prices will not necessarily follow, or at least will lag far behind the rise in asset values, resulting in an effective increase in wealth for many people—at least for a significant period.

There are, however, mitigating circumstances in regard to housing or stock asset bubbles that constrain the amount of new wealth actually achieved as "tangible services and commodities." In the case of the housing boom from 2000 to 2007, the bubble was strongly enhanced by the availability of risky marginal mortgages and the self-propagating effect of double-digit gains in asset values several years in a row. The turnover in house sales slowed remarkably in 2007, and housing prices fell dramatically. If enough people try to convert this paper profit into actual cash in 2008–2009 by selling their houses, it is likely that the CSREI will plummet further. Similarly, it seems likely that much of the 13-fold gain in stocks since 1982 has been plowed back into the stock market. If a significant number of investors start extracting cash by selling their stocks, a meltdown of the stock markets would result. Such a meltdown began in 2008. Thus, these bubbles provide a means for the nimble investor to increase his wealth by selling his assets, but if most investors were nimble, the bubble would pop.

INFLATION

As we pointed out in the previous section, a fundamental principle of Economics 101 is that supposedly, you cannot create wealth merely by raising paper asset values. Wealth is supposedly created only by increased productivity and efficiency. If the money in circulation increases but the products remain the same, you end up with more dollars chasing the same amount of goods and thus you end up with inflation. Hence you have more dollars but no real increase in purchasing power.

The baby boomer generation has set about to disprove this venerable law of economics. The baby boomers demand wealth, quick wealth, and wealth unearned. And the amazing thing is that to some considerable extent, they have succeeded—at least so far.

JKG[12] discussed inflation at some length. He began by noting that historically, inflation has mainly been fueled by war, post-war strictures, and other special situations. In the late 20th century, however, inflation has been significant during periods of prosperity. Inflation peaked in the mid-1970s. In the public arena, inflation is widely deplored and condemned by politicians of both parties, particularly conservatives. As JKG emphasized:

> Businessmen, bankers, insurance executives and nearly every type of profes sional public spokesman at one time or another have warned of the dangers of continued inflation, . . . this conviction leads to remarkably little effort and, indeed, to remarkably few suggestions for specific action. Where inflation is concerned, nearly everyone finds it convenient to confine himself to conversation.

The problem is that the remedies available are typically viewed by the Federal Reserve as being worse than the disease.

JKG provided several reasons why inflation may not be opposed by any serious effort:

a) Some people profit from inflation.

b) Some hope that inflation will eventually correct itself.

c) With the memory of the 1930s lurking in the background, most of us believe that "the most grievous threat to the American economy is a depression." That being the case, politicians are loath to take actions that could conceivably lead to an economic depression.

d) The belief is currently widespread that monetary policy by the Federal Reserve can control inflation.

Inflation can conceivably contribute to the promotion of bubbles because in times of high inflation, investors may wish to avoid holding cash, and invest in ventures that they perceive may provide protection from inflation.

However, although we have a sense of what inflation is (like wealth), it is difficult to define it exactly. Inflation is described by a website: "inflation-what-the-heck-is-it" (WTHII).[13]

WTHII provides 8 distinctly different definitions of inflation and suggests that a whole lot more may yet found. These are:

- Decline in purchasing power of the currency held.

- Rising prices in general (essentially the same as #1 although some might disagree).

- Rising consumer prices (CPI).

- Rising producer prices (PPI).

- Rising prices due to expansion of money supply.

- Rising prices due to expansion of money supply and credit.

- Expansion of money supply.

- Expansion of money supply and credit.

Unfortunately, there are several different measures of money supply, and there are consumer prices, producer prices, or simply prices in general.

Dictionary.com defined inflation as:

> A persistent increase in the level of consumer prices or a persistent decline in the purchasing power of money, caused by an increase in available currency and credit beyond the proportion of available goods and services.

Other definitions of inflation were provided by a website that specializes in inflation.[14] These range from an "increase in the amount of currency in circulation" to "a persistent increase in the level of consumer prices or a persistent decline in the purchasing power of money." Typically, some of these definitions add a phrase beginning with "because of." As WTHII pointed out, the problem with definitions that have a "because of" clause is that it is impossible to know exactly *why* prices are rising or falling.

WTHII brought up several other important aspects. One of these is the issue of whether increases in asset prices should be incorporated into indices of inflation. WTHII claimed that this would be too difficult and complicated but that does not necessarily appear to be so. Introducing stock and house prices appears to this writer to be much easier than preparing a CPI as presently defined based on a shopping cart of thousands of items. However, the real issue here is that the appropriate rate of inflation depends markedly on one's income and whether one invests in a house and securities. For poor people who own no stocks, the increase in stock market indices is irrelevant. However, the price of housing is important to the poor since it affects their ability to some day own their own home. As house prices escalate, they diverge further away from the realm where the poor could consider buying one. Hence, that may be the most important element of inflation for the poor—ultimately more important than the price of milk. Many people in the middle class typically already own their own homes, and to that extent, depending on location, may be protected from future inflation of home prices. Nevertheless, younger people who do not yet own their own home, or apartment dwellers contemplating home ownership, may be shut out of the home market by rising prices. The middle class typically depends heavily on 401(k), 403(b), and other retirement investment plans, and those in the upper range of the middle class may have (or contemplate) sizable personal investments in stocks through these retirement plans. Since the majority of people appear to believe that in the long run, stocks provide the best investment for retirement plans, many of them are committed to continuous purchase of stocks during their working years. Increasing stock prices are a benefit for older employees contemplating retirement, but are problematic for younger employees who must pay high prices for stock for many years prior to retirement. Since investments in stocks for retirement via 401(k) plans may typically constitute a major expenditure of middle class families, a rising stock market may constitute a significant form of inflation for them, particularly since such a market may be vulnerable to a future downward correction.

WTHII also introduced another issue. There are items that have become necessities (either culturally or legally) that may not have existed in the past or may have been considered to be luxuries. In either case, the cost of acquiring these goods and services adds to the cost of living as a new cost, rather than as an increase in a past cost. The example is given of double pane insulated argon gas filled windows that did not even exist 30 years ago. These are now required by code in my hometown. There are also costs for computers, cell phones, cable hook-ups, Internet service providers, and GPS navigation systems. It is not clear how these affect the CPI.

WTHII pointed out that in the 1990s, "the money supply rose dramatically by any commonly used measure" but Greenspan and the economists "were not alarmed because the price of oil and gold and copper and computers were falling." In the spirit of classical economics, WTHII then said: "Can a definition of inflation that ignores such problems possibly be right?" Apparently "such problems" refers to the increase in the money supply. WTHII then provided an explanation for how Greenspan got away with the increase in money supply without paying a price in runaway inflation: "Improvements led by an Internet revolution, along with global wage arbitrage and outsourcing to China and India, lowered costs on manufactured goods and kept the lid on wage increases in the manufacturing sector." WTHII went on to conclude:

> Those factors all helped mask rampant inflation in money supply. The Greenspan Fed further compounded the problem by injecting massive amounts of money to fight a mythical Y2K dragon that simply did not exist. Those monetary injections helped fuel a massive bubble in the stock market in 2000.

But if pumping money into the system produced a booming stock market with no price escalation, should we care if the money supply increased sharply? It appears that the 1990s represent a defeat for Economics 101: under the right circumstances it seems possible to expand the money supply, produce increases in asset values via bubbles, and not endure excessive consumer price inflation. Can this state of nirvana persist ad infinitum? The answer is probably not, because special circumstances appear to have made it possible. In fact, in the era

2002–2007, the money supply was again greatly expanded, also leading to a booming stock market, but this time the dollar weakened severely against foreign currencies and the price of oil (valued in dollars) reached new highs. The limits of the money supply expansion game appear to have been reached in July 2008.

The most widely used index of inflation is the so-called CPI-U Index (consumer price index for urban areas). The actual calculation of the index is very complex and beyond the scope of this book. Morris Rosenthal provided a brief summary.[15] He estimated that the CPI includes the following elements:

- Food and Beverages 15%

- Housing (includes utilities) 43%

- Apparel 4%

- Transportation 17%

- Medical Care 6%

- Recreation 6%

- Education and Communication 6%

- Other Goods and Services 3%

Rosenthal emphasized that the CPI-U attempts to measure monthly out-of-pocket expenses, and does not seem to take into account run-out costs. He gives the example that if everyone changed to interest-only mortgages while house prices doubled, present homeowners' costs would go down and the CPI-U would go down even while housing doubled. Rosenthal did not discuss the impact of doubling of housing on those who do not own a house, but it is devastating. The reason that medical care is listed as 6% of total expenses is that many people have medical plans through their place of employment, and they only pay for a fraction of the total cost of the plan, plus some co-payments and drugs. However, the total cost of medical care (including the cost borne by employers and Medicare) is probably more like 20% of the total and is escalating rapidly. As employers take on this rising cost, wage increases are inhibited. A recent New York Time article[16] pointed out:

Many of the 158 million people covered by employer health insurance are struggling to meet medical expenses that are much higher than they used to be — often because of some combination of higher premiums, less extensive coverage, and bigger out-of-pocket deductibles and co-payments.

A case was cited where out-of-pocket medical costs for utility workers in Arizona rose from $2,000/year to $5,600/year in five years.

Hence the CPI-U would have you believe that medical costs are a small part of the cost of living, but that is not true. Rosenthal also points out that education costs are misleading. If you have children in public school, these costs may be small for many years, but if you send them to a private college, expect a bill for about $200,000 for four years including room and board.

As the Government explanation of the CPI explains, "Purchases of houses . . . are viewed as investment expenditures and are therefore excluded" [from the CPI]. Thus, during the period 2000–2007 when house prices more than doubled across the nation and nearly tripled in some localities, there was no impact on the CPI.

A recent News Release by the Bureau of Labor Statistics (BLS)[17] indicates that according to the government:

- For the first three months of 2008, consumer prices increased at a seasonally adjusted annual rate of 3.1%. This compares with an increase of 4.1% for all of 2007.

- Transportation costs increased at a seasonally adjusted annual rate of 2.4%. There was no net change in transportation costs for February and March.

- Excluding food and energy, the cost of living increased at a seasonally adjusted annual rate of 2%.

- While the BLS does admit that energy costs went up at the rate of 8.6% (a gross underestimate), this does not seem to have affected the overall CPI very much.

Anyone who has been in the marketplace buying gasoline, food, and goods during the first half of 2008, knows this to be total fakery. For example, during February and March 2008 when gasoline prices rose steeply, the BLS claimed "there was no net change in transportation costs."

Does the conventional Consumer Price Index (CPI) have much utility? WTHII answered this question very accurately:

> "The basket of goods and services as well as subjective measures of quality improvements can indeed be used by the government to underpay holders of inflation protected securities, as well as understate cost of living adjustments to social security recipients."

SPECULATIONS, BOOTSTRAPS AND SWINDLES

Manias and bubbles typically involve borrowing large amounts of money. This may involve buying stocks on margin, borrowing from banks to finance investment, selling assets on an installment payment plan, or companies obtaining financing from speculators based on promises of future repayment. In the US real estate bubble of 2002–2007, as real estate values rose, many people refinanced their homes several times with larger mortgages, thus using their homes as a sort of ATM where you only take money out but never make deposits. For example, there was an article in the Los Angeles Times about a couple in Corona, CA who refinanced about six times from 2000 to 2007, tripling the size of their mortgage, until finally, they could not make the payments and defaulted. In this case, the banks provided funds against extrapolated future increases in real estate value as the bubble expanded. The "The Sub-Prime Real Estate Boom 2001–2007" in Chap. 2 covers the sub-prime real estate bubble in greater detail.

As K&A pointed out, the prospects for paying back these marginal or speculative loans may vary over a wide spectrum. Minsky categorized manias and bubbles according to the probability of payback on loans used to support the enterprise. In our discussion, Minsky's terminology will be changed, but his ideas are retained. Minsky distinguished between three levels of soundness of an enterprise. We will denote these as *speculations, bootstraps and swindles*.

In a *speculation*, there is a reasonable prospect that if all goes well, the operating income from the enterprise will be sufficient to pay off both the interest and amortization of its indebtedness. Barring unforeseen problems, the enterprise will be able to pay off its debt. If not, it will have to borrow to cover the amounts due on maturing loans. However, in typical cases, there are little or no reserves, so speculations are susceptible to future problems that may (and often do) arise.

In a *bootstrap* operation, it is likely that anticipated operating income will be sufficient so it can pay the interest on its indebtedness. However even making favorable assumptions, it is unlikely that the operating income will cover the amounts due on maturing loans. The enterprise must hope that it can either sell inflated stock or take on new loans to pay off the old ones as they become due. In the case of the housing bubble of 2002–2007, many people bought houses they could not afford with loans that they could not maintain past the initial favorable terms, with the expectation that if housing prices increased at 10–20% per year (as they did for several years), they could turn the house over in a year or two and make a huge percentage profit before payments rose to unaffordable levels. When housing prices stopped rising in 2007, they were unable to make the mortgage payments. If the enterprise cannot pay off loans with sale of inflated stock or by selling appreciated assets, it is doomed to continually "borrow from Peter to pay Paul" in an endless chain of borrowing to remain afloat—until they run out of lenders willing to provide them with funds.

In a *swindle*, the anticipated operating income is not likely to pay the interest or the principal on its indebtedness on the scheduled due dates; to obtain cash, the firm must continually increase its indebtedness until lenders will no longer support the venture. A *swindle* is certain to end in collapse. The epitome of a swindle is the Ponzi scheme in which one promises an inordinate rate of return to lure greedy (and gullible) investors to invest their funds in the venture. Instead of paying them interest out of earnings (of which there are none), the Ponzi scheme uses some of the capital it raised to pay fictitious interest to investors for a short period to demonstrate the supposed reality of the scheme. That lures additional investors to contribute their funds. If this were continued, pretty soon, all the invested funds would be used up paying back fictitious interest to investors. However, long before that, the operators of the scheme disappear with the remaining funds.

Minsky's hypothesis was that when the economy slows, some of the firms that had been involved in *speculation* finance are forced by circumstances into the *bootstrap* category and that some of the firms that had been involved in *bootstrap* finance group now find they are forced into the *swindle* finance group.

THE RATIONALITY OF INVESTORS?

K&A devoted a chapter to the rationality of investors and markets. Classical economics tends to attribute rationality to investors, and would therefore conclude that prices reflect information—or at least rational expectations based on currently available information. If prices rise, that would be a reflection of a rational analysis of supply/demand and future prospects that led to optimistic conclusions. But as K&A pointed out, there are many examples of irrational markets, and the nomenclature used to describe irrational markets is very diverse:

> ... manias, insane land speculation, blind passion, financial orgies, frenzies, feverish speculation, epidemic desire to get rich quick, wishful thinking, intoxicated investors, turning a blind eye, ... fools' paradise, overconfidence, over speculation, a craze . . ..

How does one explain irrational markets based on rational individual investors? K&A listed several possible explanations.

1. *Mob psychology*: In this model, virtually all of the participants in the market change their views at the same time and move as a herd.

2. *Individual crescendo*: In this model, different individuals change their market views at different stages as part of a continuing process; "most start rationally and then more of them lose contact with reality, gradually at first and then more quickly." This is well described in Ionescu's play: *Rhinoceros*.

3. *Group crescendo*: In this model, rationality differs among different groups of traders, investors and speculators, and these groups gradually succumb to hysteria as asset prices increase and the temptation to make quick profits becomes irresistible.

4. *Fallacy of composition*: This is a philosophical view that a conclusion cannot necessarily be drawn about the whole from the features of its constituents. "From time to time the behavior of the group of individuals differs from the sum of the behaviors of each of the individuals in the group." For example, an athletic team composed of outstanding players may not play well if the individuals do not integrate well.

5. *Erroneous models or information*: What appears to be irrational behavior of the group may be the result of rational behavior based on the wrong model or lack of proper information. However, that would seem to come under the category of errors in models or analysis, which though regrettable, is hardly irrational. Nevertheless, investors, working with bad models and bad information, may appear to be irrational to others.

K&A suggested that basically rational people will act irrationally at times. "Mob psychology or hysteria is well established as an occasional deviation from rational behavior." Perhaps the Schiller quote "Introduction" sums it up best:

> *Anyone taken as an individual is tolerably sensible and reasonable but as a member of a crowd, he at once becomes a blockhead.*

K&A also discussed stages of speculation. What seems to happen is that in the first phase, one invests in a venture because there is a prospect of a profit from the activities in the venture. Thus, when farm products sell for a high price, the value of farmland may rise. Initially, investors may invest in farmland with the reasonable and rational expectation that the products produced on that farmland will generate significant profits. In the second stage, as the price of farmland increases, speculators move in and buy farmland, not for the products grown on the farms, but with the intent of turning over their holdings to another speculator who may arrive on the scene later than them, having noted the expanding bubble in farmland. In the speculative stage, the original reason for investing in farmland is forgotten, and one invests only to turn over the investment to "a bigger fool." As the frenzy builds, speculators borrow to increase their leverage, and thus expand the bubble until eventually it pops.

Investors often rely on financial advice from experts. The role of investment advice was discussed by JKG in a small book entitled "Innocent Fraud." Innocent fraud is lawful fraud committed with the willing participation of the defrauded in ways that are acceptable as cultural norms. Innocent fraud is not commonly recognized to be fraud. A major example is providing financial advice by the world of finance—banking, corporate finance, the securities markets, the mutual funds, and financial guidance counselors. As JKG said, there is the inescapable fact that the future economic performance of the economy, the passage from good times to recession or depression and back, cannot be foretold. There are many predictions, and those that prove to be correct are the result more of luck than foresight. Yet as JKG pointed out, in the economic and financial worlds, there is an army of analysts who predict the unpredictable future, usually with considerable variance from reality. And many of these have rewarding careers because "what is predicted is what others wish to hear and what they wish to profit or have some return from, hope or need covers reality." JKG went on to say:

> Those employed or self-employed who tell of the future financial performance of an industry or firm, given the unpredictable but controlling influence of the larger economy, do not know *and normally do not know that they do not know.*

Nevertheless, as JKG said, "Financial advice and guidance, however worthless, can be for a time financially rewarding." And when the predictions of future glory turn out not to be true, the errors and misconceptions of experts are soon forgotten.

Typical of financial advice is the guest "expert" who appeared on "The Nightly Business Report" (NBR) on National Public Television, Friday, January 25, 2008. Apparently, he is interviewed by NBR roughly every six months. In his previous visit to NBR six months prior, he recommended three stocks. As it turned out, two of the recommended stocks from six months ago had dropped by 28% (the other rose by 8%). That dismal record posed no impediment to further prognostications by this esteemed expert, who went on to say with great confidence that the stock market had hit bottom in January, 2008 and was poised to advance from there.

Later in the same evening of Friday, January 25, 2008, a panel of editors of the Wall Street Journal presented a TV Roundtable on the Fox News Channel, in which they reiterated many times that cutting taxes (and damned be the deficit) was the key to prosperity. Their principal worry was that the Democrats, if elected, might raise the top income tax bracket from 40% to 70%, which they were certain would "wreck the economy." It is not clear to this writer how taxing the rich wrecks the economy, or whether the economy was wrecked during the 35 years (1936–1970) when the top bracket was 70% or higher.

THE RATIONALITY OF BANKERS AND EXPERTS?

RATIONALITY OF BANKERS?

If you query "Google" with the words "stupidity of bankers" you get 990,000 web pages returned.

A quote from an Internet web page is:

Intelligent banking is a contradiction in terms. Like military intelligence.

JKG[18] suggested that the wisdom attributed to financiers and bankers has made them self-satisfied and inhibits their self-scrutiny (which JKG suggests is a necessary prerequisite for "good sense"). As an example he pointed out the stupidity of New York banks and bankers that made bad loans to Latin America, Africa and Poland in the 1970s.

In 1984, Delamaide wrote a book[19] on the threat to financial stability posed by the massive lending to third world countries of petrodollars that were invested in the West by the oil-producing countries of OPEC. In Ann Crittenden's review[20] she described Delamaide's view of international bankers as "well-tailored hucksters who flew around the world selling money without any thought of the consequences." Citibank's chairman was described by Delamaide as "a glorified vacuum cleaner salesman, a small-town smooth-talker whose only goal in life was to make a buck. . ... The herd mentality that led the world banking community to follow Citibank's lead into the wilds of Africa and Latin America is presented in devastating detail, often by the bankers themselves." The banks seemed to have no idea where their money went or what it was used for. Much of it ended up as Swiss bank accounts for presidents, and in one case the American loans were used to purchase $250 million worth of military planes from Russia. Delamaide opined:

> . . . fatal flaws in the banks' foreign lending procedures—failure to obtain information on the amount of credit going to any single borrower; failure to demand collateral on the loans; failure, in short, to observe time-honored banking practices. Yet instead of moving to restrain such lending, regulators virtually everywhere accepted the banks' assurances that business should go on as usual. . .

All of this was aided and abetted by Treasury Secretary Donald Regan who was also a major figure in the Savings and Loan scandals.

Finally, Delamaide provided the following anecdote relating to the fact that as often as we go through these banking crises, we never fix the system:

> It is reminiscent of the rope basket going up to the Greek monastery. The basket was the only access to the monastery, perched on top of a mountain crag. A visitor who was about to be hauled up the sheer cliff wall noticed that the rope attached to the basket was frayed in several places. Concerned, he asked one of the accompanying monks how often they replaced the rope. 'Every time it breaks,' was the laconic response.

When one examines the savings and loan crisis of the 1980s or the sub-prime mortgage mess of the 2000s, the first questions that jump into one's mind are: Didn't bankers know this would happen? Why did they pursue obviously destructive paths? Were they dumb or crafty?

Delamaide said:

> Banks make loans, loans go bad, banks go bust. This stubbornly simple
> sequence of events was the cadence that kept bank executives, central
> bankers, and government officials dancing through the fall and spring of
> 1982–83. . . . The fear was that if a Chase Manhattan or Citibank failed, the
> shock waves would cause the collapse of many smaller banks and compa-
> nies. The greatest anxiety came from simply not knowing what would
> happen. Chase had never failed before. Who could say what the conse-
> quences would be?
>
> Kindleberger lists manias of lending to foreign countries: 1808–10, 1823–5,
> 1856–61, 1885–90, 1910–13, 1924–28. Boom loans are undertaken on the upswing,
> defaulted at the peak, and refunded in the next upswing, which may lead to new
> borrowing. In short productive loans in the developing countries are not very
> productive and do not stay long out of default. . .
>
> The supply of 'busted bonds' – those left over from 19th century defaults – is
> so great and diversified that a new hobby has grown up in the last few years of
> collecting and trading them like stamps.
>
> Latin-American countries began borrowing and defaulting as soon as they
> gained independence in the 1820s. . . . By 1940 nearly 4/5 of all Latin-American
> bonds floated in North America were in default.
>
> Commercial bankers had a lot to learn in developing countries, but their
> ignorance regarding the communist bloc was even greater.

Delamaide discussed at length the situation when the OPEC countries quadrupled
oil prices in 1973–1974, and thereby altered the world's capital flows. This created
an international imbalance in money flows with the OPEC countries collecting a
lot more money than they could spend, and the other countries were running a
deficit. The sums involved were enormous.

> The developing countries had no savings, so they had to borrow. Their deficits
> were of such magnitude that the only source of funds big enough to finance
> them were the oil surpluses themselves. But OPEC countries did not want to
> lend money directly to the countries that needed to borrow, especially not to the
> developing countries. So they deposited their money in the big international
> banks. These banks suddenly had a lot more money to lend out, but there was
> not much demand for loans in industrialized countries because of the recession
> caused by higher oil prices.

As a result, the banks started lending the money to the oil-importing developing countries that were running large balance-of-payments deficits. The OPEC countries accumulated deposits in the international banks, and the importing countries accumulated debts to the international banks. "Then the developing nations defaulted—as usual, precipitating the financial crisis of 1982–1983."

Surfing through the thousands of web pages retrieved by *Googling* "stupidity of bankers," one finds a number of recurrent themes that are popular. Some excerpts from a website blog[21] follow:

- "I doubt that bankers are idiots. While some CEOs and employees are undoubtedly stupid, I believe the subprime crisis was anticipated by the banks or some of the people in the banks. Fed and the banks are much closer than Fed and people, so they knew that the Fed would bail them out.

- Conspiracy theories or not, a lot of bankers became rich on the crisis, whether by cashing out at the right time, through insider shorting or by waiting to buy assets at dirt cheap prices. Some even made money by getting fired.

- Eventually, the crisis will be over and the banks' shares will go up. So bankers with deep pockets who have insider knowledge of Fed decisions will make more profits.

- [I don't believe] that the bankers were just blindly greedy or didn't realize what was going on over the last ten years. These guys are in the money business—They knew. It's basic economics—If you flood an economy with dollars, those dollars will end up somewhere. If you combine easy money with easy loans for homes, any moron can see that that created the bubble. We are told that, in hindsight, the banks didn't see how that worked, but I don't buy it. They knew. The heads of the banks knew exactly what was going on, and they did it anyways. Why? That I haven't figured out yet—Maybe they knew the Fed/government would come running to bail them out, so they got the money while it was good to get. Maybe the old conspiracy theory about banks creating inflation/deflation cycles to scoop up property at pennies on the dollar when things go south is true. Whatever their motive, I find it exceedingly difficult to believe that the banks simply forgot all about such elementary economic principles as bubble creation due to monetary inflation—And compounded that with all too easy loans. No, something stinks to high heaven here. . ."

Paul Krugman in the New York Times[22] asked what the Wall Street Titans were smoking when they lost staggering sums, and answered that they were high on the usual drug—greed:

> And they were encouraged to make socially destructive decisions by a system of executive compensation that should have been reformed after the Enron and WorldCom scandals, but wasn't.
>
> But even as the danger signs multiplied, Wall Street piled into bonds backed by dubious home mortgages. Most of the bad investments now shaking the financial world seem to have been made in the final frenzy of the housing bubble, or even after the bubble began to deflate.
>
> Now the bill is coming due, and almost everyone — that is, almost everyone except the people responsible — have to pay.
>
> The losses suffered by shareholders in Merrill, Citigroup, Bear Stearns and so on are the least of it. Far more important in human terms are the hundreds of thousands if not millions of American families lured into mortgage deals they didn't understand, who now face sharp increases in their payments — and, in many cases, the loss of their houses — as their interest rates reset.

Here, Krugman showed a bit of naiveté. It seems doubtful that most of those who face foreclosure did not understand the leverage they were taking on. But they, like the banks, had bubble fever and believed that 10–20% annual increases in property values would more than compensate for their debt overload. When Krugman hands out accusations of greed, he should include the millions of people who bought when they could not afford it, or upgraded when they shouldn't have, in the expectation that the inflating bubble would bail them out. As we pointed out previously, during the period 2002–2007 it was common for many people to treat their residence as an ATM machine where you only withdraw money but never make deposits.

Krugman went on to discuss collateral damage to the economy—which is substantial. However, he missed one vital point. The rise in real estate values from 2002-to-2007 created a feeling of wealth and well-being for millions of people, which encouraged them to borrow and spend, which in turn, drove the economy to prosperity which otherwise might have stagnated. The prosperity of 2002-to-2007 was certainly not due to an increase in wages.

Krugman then asked: "How did things go so wrong?" He suggested that the answers lie in (1) lack of leadership and regulation by government and the Fed, and (2) the fact that corporate executive reap huge fortunes regardless of their poor performance.

While Delamaide claimed that the bankers were foolish and shortsighted, some argue that they were fully cognizant of the risks. The nature of banking is that one can speculate with other peoples' money, make huge short-term income and bonuses, and move on when the bubble pops. And the "other people" don't lose if they have FDIC-protected accounts.

One must distinguish between banks and bankers. Banks seem to often lose huge sums of money but bankers seem to get paid well regardless of how much the banks lose. As we discuss in "Mortgage-Backed Securities" in Chap. 2, everyone made money while real estate values were going up from 2001 to early 2007. Individuals either reduced their payments through lower mortgage interest rates, or were able to borrow more and have more disposable income. Speculators were able to take ownership of expensive houses for merely the cost of the loan fee. Banks and mortgage companies earned loan fees from a very high volume of new mortgages. They promptly sold those mortgages to investment bankers, who packaged them into structured investment vehicles, which they sold to institutions and the public. However, the investment banks did not take into account:

(1) As the housing bubble expanded in 2004–2006, more and more speculators bought houses with the intent to make a quick profit before rising adjustable interest rates on their mortgage wiped them out.

(2) Banks and mortgage companies, greedy for loan fees, and believing they could simply sell off new mortgages to investment banks, increasingly granted mortgages to speculators who could not possibly qualify under conservative standards.

(3) Financial rating agencies, collecting huge profits from rating the high flux of new bonds based on mortgage collateral, were motivated to be optimistic in their appraisals, allowing bonds to be marketed to the unsuspecting.

(4) Investment banks could turn over newly issued bonds based on mortgage collateral almost as fast as they were produced.

Where banks and mortgage companies went wrong, was in holding on to too many of their sub-prime mortgages, instead of getting rid of them to investment banks. Where investment banks went wrong was in not getting rid of the mortgage-backed bonds quickly enough, and holding some for their own accounts. Where the public went wrong was in believing the ratings on mortgage-backed bonds. While the ultimate toll from all this is likely to exceed a trillion dollars, the new wealth created out of thin air from the housing bubble was over $5 trillion (see "Residences as ATMs" in Chap. 2).

RATIONALITY OF EXPERTS?

The books by JKG are replete with reports of specious and unfounded prognostications by commercial magnates during the boom years of the 1920s that supported speculation in stocks. Many of these experts "assured . . . that [the stock market bubble] was well within the norms of contemporary and successful capitalism." JKG quoted Irving Fisher of Yale. . . . who "gained enduring fame for the widely reported conclusion that stock prices have reached what looks like a permanently high plateau."

Only a few took exception (Paul M. Warburg and Roger Babson) but they were widely condemned. JKG traced out the history of the Harvard Economic Society that abandoned its summer position of pessimism late in 1929. In a series of quotations from November 1929 through 1930 and on into 1931, this learned economics society continued to sound the message of optimism. In November it said: "a serious depression like that of 1920–1921 is outside the range of probability." In December 1929, its forecast for 1930 was "A depression seems improbable; [we expect] recovery of business next spring, with further improvement in the fall." It repeated this judgment on November 23, 1929 and on December 21, 1929. The phrases in their reports over the next two years included: ". . . the severest phase of the recession is over," "definitely on the road to recovery," "the outlook continues favorable," and many more prognostications of the same ilk. As JKG pointed out finally:

> Somewhat later, its reputation for infallibility rather dimmed, the Society was dissolved. Harvard economics professors ceased forecasting the future and again donned their *accustomed* garb of humility.

Professor Irving Fisher of Yale was a leading economist of his time who had authored books with titles "The Purchasing Power of Money," "The Rate of Interest," and "The Theory of Interest." One of his arguments rested on the benefits he saw flowing from prohibition, citing the work of Columbia Professor Paul Nystrom, who concluded that "a dry nation would increase the efficiency of workers and switch demand from liquor to home furnishings, automobiles, musical instruments, radio, travel, amusements, insurance, education, books and magazines." He produced a continuous stream of theories, books, and speeches during the 1920s supporting the expansion of the stock market bubble as being well grounded in solid fundamentals. A few days before the Stock Market Crash of 1929, Fisher insisted:

Stock prices have reached what looks like a permanently high plateau. I do not feel there will be soon if ever a 50 or 60 point break from present levels, such as (bears) have predicted. I expect to see the stock market a good deal higher than it is today within a few months.

At the first break in the 1929 markets, he insisted that the market was "only shaking out of the lunatic fringe" and went on to explain why he felt the prices still had not caught up with their real value and should go much higher. The New York Times, Oct. 22, 1929 quoted him as saying: "Security values in most instances were not inflated... The nation is marching along a permanently high plateau of prosperity... Any fears that the price level of stocks might go down to where it was in 1923 or earlier are not justified by present economic conditions." For months after the crash, he continued to assure investors that a recovery was just around the corner. JKG summarized:

Professor Irving Fisher tried hard to explain why he had been wrong. Early in November 1929 he suggested that the whole thing had been irrational and hence beyond prediction. In a statement that was not a model of coherence, he said: *It was the psychology of panic. It was mob psychology, and it was not, primarily, that the price level of the market was unsoundly high . . . the fall in the market was very largely due to the psychology by which it went down because it went down.*[23]

The explanation attracted little attention except from the editor of The Commercial and Financial Chronicle. The latter observed with succinct brutality: "The learned professor is wrong as he usually is when he talks about the stock market." The mob he added, "didn't sell. It got sold out."

Before the year was over, Professor Fisher tried again in his book, "The Stock Market Crash—and After." He argued, and rightly for the moment, that stocks were "still on a plateau, albeit a somewhat lower one than before, that the crash was a great accident," that the market had gone up "principally because of sound, justified expectations of earnings." He also argued that prohibition was still a strong force for higher business productivity and profits, and concluded that for "the immediate future, at least, the outlook is bright." This book attracted little attention. As JKG emphasized: "One trouble with being wrong is that it robs the prophet of his audience when he most needs it to explain why."

For many years, Fisher was regarded as a pariah of the 1920s stock market bubble, but in recent years, his personal stock has risen. A Federal Reserve Report[24] was published in December, 2003 entitled "The 1929 Stock Market: Irving Fisher Was Right." The Abstract of this Fed Report is:

Many stock market analysts think that in 1929, at the time of the crash, stocks were overvalued. Irving Fisher argued just before the crash that fundamentals were strong and the stock market was undervalued. In this paper, we use growth theory to estimate the fundamental value of corporate equity and compare it to actual stock valuations. Our estimate is based on values of productive corporate capital, both tangible and intangible, and tax rates on corporate income and distributions. The evidence strongly suggests that Fisher was right. Even at the 1929 peak, ***stocks were undervalued relative to the prediction of theory***.

It may be true that stocks were undervalued in 1929 relative to an abstract theory of economics, but if one considers a more pragmatic view that stocks were worth what investors were willing to pay for them, then they were clearly and massively overvalued.

The Fed Report went on to ask: "If stock prices were not inflated beyond their fundamental values in October 1929, why did the market crash?" They suggested that tightening of money by the Federal Reserve was the principal cause.

However, JKG[25] provided a lengthy discussion of the role (such as it was) of the Federal Reserve during the boom years of 1928–1929. He described the Fed in those days as being "not so much unaware or unwilling, but impotent." He demonstrated that the speculative fever was so intense that any rise in interest rates would be no deterrent. Nevertheless, the Fed vacillated for six months after March 1929 before finally raising the discount rate from 5% to 6% in August 1929. The stock market's response was a yawn, and it accelerated upward. JKG went on to assert:

> The collapse in the stock market in the autumn of 1929 was implicit in the speculation that went before. The only question regarding that speculation was how long it would last.

It is worth noting that learned economists have continued to discuss (one way or the other) whether a "bubble" existed in stock prices in the 1920s. G. J. Santoni (of the Fed) argued against a bubble (see subsequent pages). White cautiously suggested that it was likely that there was a bubble.[26] De Long and Shleifer[27] concluded "that a substantial component of the rise in stock prices up to and fall of stock prices away from September of 1929 was in fact excessive, and not based on rational revisions of warranted valuations."

A book was published in 1998 that purported to analyze possible contributing causes for the crash of 1929, and to some extent absolved Irving Fisher.[28] The possible causes for the 1929 crash were listed as:

1. The stock market was too high in September 1929 (values did not justify prices) because of excessive speculation and the crash was inevitable.

2. A real downturn in business activity.

3. The Hatry affair in England and the subsequent raising of interest rates in London, and liquidation of English investments in the United States.

4. Actions of the Federal Reserve Board.

5. The message being sounded by the media and by important governmental figures on both sides of the Atlantic that the US stock market was too high and there was a "war" against the speculators.

6. Excessive buying on margin and excessive buying of investment trusts.

7. Excessive leverage when the debt of operating utilities, holding companies, investment trusts, and margin buying are all considered.

8. The setback in the public utility market arising from an adverse decision for utilities in Massachusetts (refusal to allow splitting of stock to encourage investment).

9. Over-reaction by the market.

The flyer for his book says:

> Attempting to reveal the real causes of the 1929 stock market crash, Bierman refutes the popular belief that wild speculation had excessively driven up stock market prices and resulted in the crash. Although he acknowledges some prices of stocks such as utilities and banks were overpriced, reasonable explanations exist for the level and increase of all other securities stock prices. Indeed, if stocks were overpriced in 1929, then they more even more overpriced in the current era (1998) of staggering growth in stock prices and investment in securities.[29] The causes of the 1929 crash, Bierman argued, lie in an unfavorable decision by the Massachusetts Department of Public Utilities coupled with the popular practice known as debt leverage in the 1920s corporate and investment arena.

Bierman rejected the first two causes. He believed that elements 6 through 9 triggered the crash, and element 3 contributed to it. He concluded:

> The overall stock market was not excessively high in September 1929 and the business outlook was favorable. Thus, the October crash did not occur because the market was too high. However, at least one segment of the market (public utilities) was too high and too leveraged, and the stage was set for the selling panic by the press and governmental officials repeatedly speaking of an orgy of speculation.

Note that Bierman, unlike McGrattan and Prescott, did not think that Fed actions were a major factor in the crash. It is also noteworthy that Bierman did not seem to think that a quadrupling of stock prices in a few years qualified as an "orgy of speculation."

JKG took a very different viewpoint. He emphasized that the economy [rather than the stock market] reached a peak in June 1929 and "then turned down and continued to decline throughout the rest of the year. . . .. Production . . . for the moment, had outrun consumer and investment demand for them." He suggested that in 1929 "modifying a famous cliché, the economy was fundamentally unsound." He provided five factors underlying this weakness: (1) distorted distribution of income amongst Americans, (2) poor corporate structure, (3) poor banking structure, (4) problems with foreign exchange, and (5) bad economic intelligence. These themes are elaborated in "The Roaring 20s Stock Market," "The Great Depression of the 1930s" in Chap. 2.

A senior economist at the Federal Reserve Bank of St Louis wrote a learned paper in 1987 disputing the attribution of stock market crashes to bursting of bubbles.[30] G. J. Santoni emphasized that "Many people attribute the bull markets of 1924–1929 and 1982–1987 and the subsequent collapses to speculative bubbles in which a crash was inevitable until the bubble burst" and he showed the similarity of these two bull markets with a graph similar to Figure 1.3.

Santoni quoted many experts who blamed these bull markets and collapses on "gambling, widespread intense optimism, overpriced due to speculation, absurdly high stock prices, unjustifiably high prices of common stocks, greed and fear, and that the collapse of 1929 was implicit in the speculation that went before."[31]

However, Santoni went on to say that "if stock price bubbles exist, economic policy makers face a difficult problem because bubbles suggest that plans to save and invest may be based on irrational criteria and subject to erratic change." This seems to imply that if the behavior of the investing public is

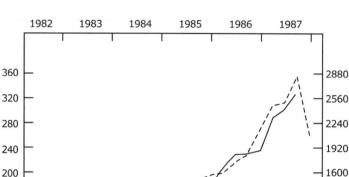

Figure 1.3 Comparison of bull markets of 1924–1929 and 1982–1987. (Adapted from: "The Great Bull Markets 1924–1929 and 1982–1987: Speculative Bubbles or Economic Fundamentals?" by G. J. Santoni).

beyond rational prediction, then the Fed is in a quandary as to how to react. But the pejorative tone of the sentence suggests significant doubt that bubbles really do occur, despite the evidence of Figure 1.3, for example. Santoni went on to say:

> If stock price bubbles exist, the periods 1924–29 and 1982–87 are likely places to look for them.

This is a very reasonable conclusion. On the other hand, he investigated the possibilities that these increases in stock prices may have been in line with economic fundamentals—whatever that means. Santoni's complaint seems to have been that there is no economic mathematical formula that can be used decisively to determine whether a bubble is occurring or has occurred. He demanded specific mathematical criteria by which to judge whether in fact a bubble has formed and claimed that attributions of bubbles are made subjectively and that "attributing crashes in stock prices to bursting bubbles adds nothing to our understanding of why crashes occur or how to prevent similar occurrences in the future." Finally, Santoni concluded:

This paper provides evidence contrary to the notion that the crashes were the result of bursting bubbles. Rather, the data suggest that stock prices followed a random walk.

Mr. Santoni did not seem to be bothered at all by sequentially repeated $\sim$30% increases per year in stock prices over a several-year period. Such stock price increases seemed to be neither incredible, nor unusual, nor unreasonable to him. And if such increases are indeed in line with economic fundamentals, that would seem to imply that either (1) at the beginning of such a multi-year sequence stocks were grossly underpriced, or (2) that economic growth at 30% per year is commonplace. In fact, no price rise in paper assets seems to have outraged Mr. Santoni's sense of proportion. The only thing that bothered him was that "crashes occur" and they need to be prevented. In other words, the end product of repeated 30% yearly gains needs to be preserved. If Mr. Santoni could figure out a way of doing that, he would make a vital contribution, and maybe we would all be rich. Finally, it would seem most strange that a random walk would produce a repetitive pattern of the sort shown in Figure 1.3.

Economists, and particularly those who are employed by the Federal Reserve have difficulty identifying and characterizing bubbles. Part of their problem seems to be that they are so encased in economic theory that they don't necessarily observe what is happening around them.[32]

MONETARY POLICY AND THE FEDERAL RESERVE SYSTEM

Central Banks, such as the US Federal Reserve, are supposed to be politically independent and are charged with the responsibility to maintain liquidity in the banking system, while avoiding excessive inflation.

As JKG pointed out, the two classic instruments that the Federal Reserve has in dealing with bubbles are open market operations and manipulation of the rediscount rate. Open market sales of governmental securities removes money that otherwise might be used for speculation, and would tend put a damper on speculation. Theoretically, increasing the discount rate would tend to discourage speculation with borrowed dollars because the cost of borrowing would increase. However, JKG also pointed out that an increase of say 1% in the rediscount rate can hardly discourage an investor during boom times who believes he will make tens or hundreds percent gains with the borrowed money. During the stock bubble of the late 1920s, the interest rate for purchasing stocks on margin sometimes reached as high as 20%, but this did not discourage investors.

Since about the 1980s, the investment public's faith in the effectiveness of monetary policy in fulfilling its charter to maintain liquidity and control inflation seems to have grown by leaps and bounds. Each hint, innuendo or implication in Federal Reserve notes and releases is given great weight and has an immediate short-term effect on the stock markets. Richard Fisher, Dallas Federal Reserve President was quoted as saying:

> Think of the Fed funds rate as a monetary spigot, and the Fed's goal is keeping the lawn of the economy green and healthy. If we turn the spigot up too forcefully, we will flood and kill the grass with inflation. If we provide too little, the lawn turns brown, starved for money.

However, the reality of the complexities of the investment banking sector led Fisher to admit that:

> ... even as we have been cutting the fed funds rate, even as we have been opening the monetary spigot, interest rates for private sector borrowers have not fallen correspondingly, and rates for some borrowers have increased. The grass is turning brown.

Or, as Mr. LaMonica put it:

> In other words, the rate cuts have not substantially helped and more big rate reductions might not either. Instead, they may simply lead to a further weakening of the dollar and more inflation.[33]

In late 2007 and early 2008, hints and innuendos regarding possible future rate cuts by the Fed made financial headlines and spurred the stock markets to sudden rallies. Fortunately for speculators, the Fed always came through with rate cuts in 2007–2008 whenever the stock market faltered. The stock market faltered in March 1929, and according to JKG, "March 26, 1929 could have been the end" [of the bubble, had the Fed clamped down on the money supply]. However the Fed, when faced with the choice of popping the bubble or prolonging it, chose the latter course.

While theoretically, one may think that part of the responsibility of the Fed is to stabilize the economy by exercising constraint over budding bubbles through monetary policy, in actual practice the Fed is beholden to the current political administration which desires to be returned to office at the next election. History amply demonstrates that people are much happier when their assets are growing by leaps and bounds, than they are when the Fed constrains the money supply to keep the economy in check. As a result, the Fed has actively intervened many times after 1990 to prop up faltering asset markets and perhaps unwittingly, support expansion of bubbles. The stock markets made incredible gains from 1995 to early 2000. Table 1.2 shows the gains made by the markets prior to the crash that started in 2000. Alarmed at the expanding bubble, Arthur Greenspan made his now famous comment "irrational exuberance" in late 1996. However, there is no evidence that the Fed took any action against this bubble. After 1996, there were no further negative comments by Greenspan, but he did make quite a few rationalizations for the expanding bubble. Apparently, Greenspan desired to stay in office and be well regarded. The best way to do this was to allow as many people as possible to get rich quickly from asset growth.

Along the way, the Fed engineered a bailout for the defunct LTCM hedge fund in 1998 (see "Long-Term Capital Management" in Chap. 2).

Artificially low interest rates create an illusion of wealth that tempts borrowers into taking on unsustainable debt. This illusion produces a temporary excess of demand over supply, driving up sales. Easy loans make expensive consumer goods suddenly appear affordable. Low interest rates and excessive money supply consequently cause asset prices to rise. During the real estate bubble from 2002 to 2007, the Fed drove down interest rates, reducing the national savings rate to negative values. No constraints were placed on banks issuing very risky mortgages, and a housing bubble resulted.

However, JKG was highly critical of monetary policy as an instrument of financial management. Thus, he said:

There is no magic in the monetary system, however brilliantly or esoterically administered, which can reconcile price stability with the imperatives, of

Table 1.2 Gains of the S&P 500 and NASDQ during 1995–2002

Year	Gains of the S&P 500 (%)	Gains of the NASDQ (%)
1995	+37.4	+39.9
1996	+23.1	+22.7
1997	+33.4	+21.6
1998	+28.6	+39.6
1999	+21.0	+85.6
2000	−9.1	−39.3
2001	−11.9	−21.1
2002	−22.1	−31.5

production and employment as they are regarded in the affluent society. On the contrary, monetary policy is a blunt, unreliable, discriminatory and somewhat dangerous instrument of economic control. It survives in esteem partly because so few understand it, including [those] on whom it places the prime burden of its restraint.

As we pointed out earlier, Robert E. Lucas (Nobel laureate in economics) argued against the common belief that easy money policy with low interest rates boosts economic growth, suggesting that any attempt to boost growth through reducing interest rates is counterproductive.

On January 17, 2008, Richard W. Fisher, head of the Dallas branch of the Federal Reserve System, gave a speech in which he elaborated his views on the role of the Fed.[34] In this speech, he asserted:

Our job is not to bail out imprudent decision makers or errant bankers, nor is it to directly support the stock market or to somehow make whole those money managers, financial engineers and real estate speculators who got it wrong. And it most definitely is not to err on the side of Wall Street at the expense of Main Street.

These are good words indeed, but are they credible? The day after the crash of October 19, 1987, the Fed acted to provide liquidity to the financial system *"in an effort to restrain the declines in financial markets* and to prevent any spillovers to the real economy."[35] In late 2007 and early 2008, the Fed reacted promptly to every downturn in the stock market with a rate cut. In 2008–9 the Treasury Department spent over a trillion dollars to bail out money managers "who got it wrong."

Fisher said that the Fed operates under two mandates: "grow employment *and* contain inflation." He then went on to assert: "the Fed has delivered on its mandate."

One cannot help but wonder whether the Fed would take credit for the sunrise every morning by facing east and saying: "arise oh sun." Inflation has indeed been under control, but mainly because real wages have been rather stagnant for 25 years, and the main way that people have become more prosperous is through inflated stock and real estate bubbles.

Mr. Fisher then attacked inflation as a scourge that "ultimately proves debilitating for businesses, consumers, investors . . . and especially for the poor, the elderly and people on fixed incomes." He said that inflation "inculcates bad financial behavioral patterns in the young by encouraging spending rather than investment and saving. Inflation is bad for Main Street and Wall Street." He reaffirmed his dedication to combat inflation. However,

I have observed that what is worst for the poor, the elderly and people on fixed incomes on Main Street is low interest rates, thus robbing them of a meager income from their savings. The Fed has proved itself willing and able to slash interest rates to preserve bubbles; how can it claim to desire to support Main Street and encourage saving?

William Poole, President of the Federal Reserve Bank of St. Louis presented a speech entitled: "Real Estate in the US Economy" before the Industrial Asset Management Council Convention in St. Louis on October 9, 2007. His speech seemed to minimize the depth of the financial problem facing the country in the wake of the punctured real estate bubble. He said: "Unfortunately, recent events suggest that housing will remain weak for several more quarters; stabilization may not begin until well into 2008." That has proven to be a major understatement. It appears that the Federal Reserve grossly underestimated the extent and depth of the aftermath of the puncture of the real estate bubble.

What is more instructive, however, is Poole's statement:

The Federal Reserve has neither the power nor the desire to bail out bad investments. We do have the responsibility to do what we can to maintain normal financial market processes. What that means, in my view, is that we want to see restoration of active trading in assets of all sorts and in all risk classes. It is for the market to judge whether securities backed by sub-prime mortgages are worth 20 cents on the dollar, or 50 cents, or 100 cents. Obviously, the market will judge different sub-prime assets differently, based on careful analysis of the underlying mortgages. That process will take time, as it is expensive to conduct the analysis that good mortgage underwriting would have conducted in the first place. Although there is a substantial distance to go, restoration of normal spreads and trading activity appears to be under way, and we can be confident that in time the market will straighten out the problems. We do not know, however, how much time will be required for us to be able to say that the current episode is over.

After observing the Federal Reserve's panicky attempt to thwart the inevitable collapse of the real estate bubble via a succession of rate cuts from November 2007 to April 2008, we now have *prima facie* evidence that, contrary to Poole's assertion, the Fed does have the **desire** to bail out bad investments. Whether it has the **power** to bail out bad investments remains to be seen.

We have previously quoted a Fed official who said that after the crash of 1987, the Fed acted to **"restrain the declines in financial markets."**

The exact role of the Fed in "restraining the declines in financial markets" remains uncertain. A provocative report[36] suggested that the Fed's role in protecting declining asset markets throughout the past few decades has been more active and direct than is generally realized. However, the report relied upon innuendo, inference and reading between the lines of statements by high officials; there is no firm evidence for any of the claims that were made. Nevertheless, the views expressed in this report are not beyond credibility and are likely to represent fact to some considerable extent. Here is a brief digest of their conclusions:

- The United States has a so-called "Plunge Protection Team" whose primary responsibility is the prevention of destabilizing stock market declines. Comprising key government agencies, stock exchanges and large Wall Street firms, this informal group was apparently created in 1989 as an outgrowth of the President's Working Group on Financial Markets.

- At the time of the Long-Term Capital Management crisis in 1998, the Federal Reserve directed large banks to prop up the currency markets. This was apparently done to diffuse a global currency crisis.

- In response to the September 11 terrorist attacks, the Federal Reserve and large Wall Street firms prepared to support the main stock markets by buying shares if panic selling ensued. Investment banks and brokerage houses took concerted actions in the aftermath of the tragedy.

- Before the 2003 Iraq invasion, the United States and Japan reached an agreement to intervene in stock markets if a financial crisis occurred during the war. Though it was announced at a press conference by a Japanese government official, the United States never publicly acknowledged the accord.

- The stability of domestic stock markets is considered by the US government to be a matter of national security. Interventions are likely justified on the grounds that the health of the US financial markets is integral to American preeminence and world stability.[37]

- A 1989 "USA Today" story revealed that Government regulators asked market participants to buy stocks in October 1989 to prevent another plunge. When these overtures proved ineffective, large brokerage firms appear to have intervened in the futures market to support the underlying index. In this regard, the recovery was remarkably similar to the miraculous turnaround in equities the day following the 1987 crash.

- The Fed will attempt to stabilize plunging stock markets by purchasing stock index futures contracts. Such a move would force the underlying index to rise. There are implications that the government supported the stock market in 1987, 1989 and 1992.

Even an astute observer like Shiller can be misled.[38] He argued that credit tightening was an important contributor to the crash of 1929 and the ensuing depression. He said:

> There have been occasions on which tightened monetary policy was associated with the bursting of stock market bubbles. For example, on February 14, 1929, the Federal Reserve raised the rediscount rate from 5% to 6% for the ostensible purpose of checking speculation. In the early 1930s, the Fed continued the tight monetary policy and saw the initial stock market downturn evolve into the deepest stock market decline ever, and a recession into the most serious US depression ever.

However, as JKG showed, nothing could be further from the truth. First of all, the stock market inflated unabated after the February 1929 increase in the discount rate. The stock market shrugged off the increase in the interest rate. Second, the bubble mentality was so frothing that investors were happy to pay double-digit margin interest rates to plough more money back into the stock market. What might have contributed more to the demise of the bubble and formation of the depression was fiscal policy in which taxes were raised to balance the budget—which was what JKG calls the "conventional wisdom" of the times.

Shiller also believed that when Japan raised the discount rate from 2.5% to 6% at the peak of the Japanese stock market between May 1989 and August 1990 "which were thought to have become overpriced because of easy monetary policy . . . this action by the bank played some role in the stock market crash and severe recession that followed." In saying this, he seems to imply that all was well in Japan with the Nikkei at 35,000 and the bubble would have endured had it not been for a tightening of credit. He does not seem to consider that the Japanese bubble was greatly over-inflated with the Nikkei at 35,000 and was likely to pop of its own accord had it been left to run its course.

Shiller asserted that tightening money "has the potential to exert a devastating impact on the economy as a whole" but may not strongly affect the expansion of a bubble. He concluded:

A small, but symbolic, increase in interest rates by monetary authorities at a time when markets are perceived by them to be overpriced is a useful step, if the increase is accompanied by a public statement that it is intended to restrain speculation. But authorities should not generally try to burst a bubble through aggressive tightening of monetary policy.

He did not discuss the counterpoint to this. Should the monetary authorities merely make a small, but symbolic, *de*crease in interest rates at a time when markets are perceived by them to be falling rapidly? In retrospect, it now seems clear that decreased interest rates do not stimulate business directly as claimed by many economists. Instead, the major effects of reduced interest rates are: (1) increased borrowing, (2) depreciation of the dollar against foreign currencies and a rise in the international price of oil, (3) discouraging savings and investments in interest bearing securities, and (4) promotion of bubbles in paper assets.

> While the common belief is that reduced interest rates stimulate business by reducing the cost of borrowing, it seems likely that the major effect of reduced interest rates is mainly to discourage saving, and the funds that would have gone into saving then migrate into paper assets: stocks or real estate. The valuation of these assets rises as demand increases, producing what Mr. Greenspan called the "wealth effect" in which people feel wealthier, spend more, and stimulate the economy. Lower capital gains taxes also contribute to such booms. Hence, the stimulus for business produced by lower interest rates is mainly a consequence of bubble formation.

The Federal Reserve System apparently believes that there is no end to American borrowing, and is oblivious to the fall of the dollar or the rise in the price of oil. The Federal Reserve System serves the rich who profit the most from bubbles, and cares not for the loss of income to savers.

JKG provided his assessment of the Federal Reserve System[39] when he described it as "our most prestigious form of fraud, our most elegant escape from reality." JKG began by emphasizing the lofty regard in which the Federal Reserve (and more specifically, its long time chairman, Mr. Greenspan) is held for supposedly controlling inflation and recession. As JKG said:

Quiet measures enforced by the Federal Reserve are ... manifestly ineffective. They do not accomplish what they are presumed to accomplish. Recession and unemployment or boom and inflation continue. Here is our most cherished and, on examination, most evident form of fraud.

JKG's claim was that all recessions eventually come to an end, but not because of actions taken by the Fed. However, the mystique of Greenspan and the Fed is so pervasive that the Fed "will receive credit if and when there is full recovery." But JKG insisted:

When times are good, higher interest rates do not slow business investment. They do not much matter; the larger prospect for profit is what counts. And in recession or depression, the controlling factor is the poor earnings prospect. At the lower interest rates, housing mortgages are refinanced; the total amount of money so released to debtors is relatively small and some may be saved. Widespread economic effect is absent or insignificant.

However, as the sub-prime mortgage fiasco indicates, if twelve trillion dollars worth of mortgages are refinanced, a housing bubble may result. JKG seems to have underestimated the potential for such a housing bubble resulting from low interest rates.

Finally, JKG described the Federal Reserve in 1929 as "a body of startling incompetence." Is there any reason to believe that anything has changed since then?

FISCAL POLICY AND TAXES

TAX POLICIES

INCOME TAX

While monetary policy is concerned with the money supply and interest rates, fiscal policy is concerned with the balance between government spending and tax receipts, producing either a budget surplus or a budget deficit. According to JKG, liberals tend to prefer fiscal policy, whereas conservatives tend to prefer monetary policy in managing the economy. The effect of fiscal policy is more easily described than the effect of monetary policy. Fiscal policy primarily affects consumer spending whereas monetary policy primarily affects business investment.

However, JKG believed that fiscal policy has not proven itself to be a good defense against inflation.

In the post-WWII years, the conservatives in the United States preferred to constrain government spending and taxes, and sought the holy grail of a balanced budget. Constrained government spending implied minimal support of social welfare programs, and as a result, the conservatives were described as heartless by their opposition. By contrast, liberals were more concerned with providing for the needy, and in order to secure funds for social welfare programs, they were willing to raise taxes, particularly on the rich (at least in theory if not in practice). As a result, the opposition to liberals described them as spendthrift and labeled them as advocates of a "tax and spend" policy.

The Republican Party has had a strong aversion to increased taxes for a considerable time. Unfortunately, as government spending has inevitably increased, particularly through "entitlements," this has made it difficult, if not impossible, to produce a balanced budget. In addition, the high cost of defense, strongly endorsed by Republicans, which produces no products of use to the public, continues to increase. Hence the two essential elements of conservatism: a balanced budget and low taxes have proven to be incompatible. When push comes to shove, the Republicans have been willing to forego a balanced budget, but treat a tax increase as a veritable policy from hell delivered by the devil. As a result, the Bush administrations of 2000–2008 have generated unprecedented high budget deficits through tax cuts. This was intended to "correct" the "mistake" that George Bush senior made when he raised taxes slightly during his 1988 term. In July 2008, president Bush announced that the yearly deficit for 2008 would hit a new high.

One of the amazing things that is difficult to comprehend, is the fact that the Republicans have swayed the voters into accepting their tax policies, which are contrary to the interests of all but the rich. These policies involve the following machinations:

(a) The Republicans enact an across-the-board tax reduction that produces huge tax cuts for the rich and very modest tax cuts for the middle class. This may be disguised as an across-the-board cut of equal percentage, say 2% to all. But 2% of $50,000 is a thousand dollars, whereas 2% of $50,000,000 is a million dollars. Alternatively, it could be a cut in capital gains taxes, and since the rich make a much higher proportion of their income from capital gains, this benefits the rich the most by far. Thus the Republicans throw the middle class a bone, and eat the roast themselves. And the public loves them for it.

(b) If the Democrats propose to modify the tax code by reducing taxes on the poor and increasing taxes on the rich, the Republicans blare in stentorian tones: "The Democrats want to raise taxes!" And the public seems to fall for it every time.

The Bush administration passed tax cuts in 2001 and 2003 that mainly benefited the rich and led to huge budget deficits. One of the strange things about these tax cuts is they expire at the end of 2010. But it is not immediately clear why they should expire at all, although the answer seems to be linked to the ensuing budget deficits in the aftermath of the tax cut, and the perceived propaganda value of limiting the cumulative deficit produced by the tax cuts by limiting the tenure of the cuts. However, the Republicans had control of Congress, so it is not clear why they needed to do this. Furthermore, since they were evidently expecting deficits, how could they claim that reduced taxes would reduce deficits? That is purely a rhetorical question. As we approach 2010, with the likely prospect of a Democratic administration in that year, the future of these cuts remains in doubt. Two strange aspects of the possible expiration of the tax cuts arise. One is that with the prospect of higher capital gains tax in 2011, 2010 might prove to be a very good year to take capital gains by selling assets. Thus all the asset bubbles may pop in 2010. Another aspect is that the inheritance tax is programmed to go to zero in 2010, but return to its longstanding level of about 50% with only a modest exemption in 2011. Hence, by committing suicide in 2010, a sickly person can double the legacy left to his heirs compared to dying in 2011. *In other words, it pays to die in 2010.*

The maximum income tax bracket in the United States has historically been much higher than it is today. From 1917 through 1986, a period of 70 years, the maximum income tax rate in the highest bracket was 50% or higher, except for the interval 1925–1931 when it was 25%. The top tax rate was over 90% from 1944 to 1963, and was 70% or higher from 1936 through 1970. The first thing that President Reagan did on taking office in the 1980s was to reduce the maximum income tax rate from 70% to 50%. In 1987, he strove to further cut this rate to 28%. David Cay Johnston (DCJ) suggests that there was an evangelical drive to do this, as if it would bring ruin on the nation if the maximum rate were as high say, 29%. However, with the Democrats controlling both houses of Congress, in order to push this second tax reduction for the rich through Congress, Reagan's administration needed to claim that the new tax system would generate as much revenue as the previous one. To achieve this, they made some changes to the tax laws, including:

- Expansion of the definition of taxable income.

- Elimination of many tax shelters

- Adding personal exemptions and the standard deduction to the AMT. (Note that the AMT was originally devised to prevent the super rich from avoiding

49

taxes, whereas personal exemptions and the standard deduction are of greatest use and value to the poor and the middle class. This policy, endorsed by the Democrats, was a direct attack on the poor and the middle class, lowering the income threshold at which taxpayers began paying the AMT.)

- Creating a "bubble" in the income tax bracket structure whereby income in the range $71,900–$149,250 was taxed at 31%, while income greater than $149,250 was taxed at only 28%. (This was the first—and hopefully last—time that such an inversion of tax rates was ever enacted, and it was done with full complicity of the Democrats.)

Two bizarre aspects of the tax revision of 1987 were:

Although the regular income tax rates were indexed for future inflation, the AMT was not. As the years went by, this led to the problem that millions of taxpayers became susceptible to paying the AMT.

Despite the supposed removal of tax shelters, in 1987, 140,000 taxpayers paid the AMT but 472 very high gross incomes paid no taxes at all.

There are a great number of websites on the Internet, and learned papers by economists, often in the employ of right wing "think-tanks," that explain why a policy of "soaking the rich" is unproductive and self-defeating. Fox News broadcasts this message daily. Most of these websites claim that the rich are already paying an inordinate share of income taxes, and that there is not much more that can be squeezed out of the rich. These websites emphasize Row 1 in Table 1.3, showing that the top 20% of households contribute 86.3% of income taxes, and the top 5% contributes 60.7% of total income taxes. These data would suggest that the rich are already shouldering a very heavy burden, and it would be grossly unfair to ask more of them. However, as the sayings go, "figures don't lie but liars do figure" or "there are lies, damned lies and there are statistics." The percentage of total income tax is a meaningless figure by itself. As Row 2 of Table 1.3 shows, the top 20% of households account for 55.1% of total income in the United States. More importantly, as Row 7 of Table 1.3 shows, the top 20% of households own 84.6% of the wealth in the United States and the top 5% own 58.9% of the wealth. Thus, the percent of income taxes paid by the upper strata is very much in line with their wealth. This data is not displayed by the defenders of the rich.

Note that the average income of those in the top 20% is 15 times higher than those in the bottom 20%. Those in the top 1% have incomes 1000 times higher than those in the bottom 20%. Another interesting statistic is line 6 of Table 1.3, the sum of effective income, SS and Medicare tax rates. This rate is about 21% for the uppermost 20% of households. There appears to be plenty

Table 1.3 Who pays how much income tax in the United States?
(2005—from Congressional Budget Office)

Households	Lowest 20%	Second 20%	Third 20%	Fourth 20%	Top 20%	Top 10%	Top 5%	Top 1%
1. Percent of total income taxes paid	−2.9	−0.9	4.4	13.1	86.3	72.7	60.7	38.8
2. Pre-tax income (% of total)	4.0	8.5	13.3	19.8	55.1	40.9	31.1	18.1
3. Average income per household ($1000s)	15.9	37.4	58.5	85.2	231	339	520	1560
4. Effective federal income tax rate (%)	−6.5	−1.0	3.0	6.0	14.1	16.0	17.6	19.4
5. Effective SS and Medicare tax rate (%)	8.3	9.2	9.5	9.7	6.0	4.8	3.5	1.7
6. Sum of effective income, SS and Medicare rate (%)	1.8	8.2	12.5	15.7	20.1	20.8	21.1	21.1
7. Percent of total wealth in United States	0.0	0.2	3.8	11.3	84.6	71.2	58.9	34.3

of room to raise this rate. If the 21% rate for the top 20% of households was increased to say, 31%, the gain in income for the federal government would be 10% x $231,300 x 23,000,000 households = $530 billion, wiping out the federal deficit.

ESTATE TAX

For some considerable time, there has been a tax of approximately 55% on estates above an exemption amount that stood at $675,000 in 2001 (double that for couples). The Republicans have campaigned against the estate tax for many years, calling it a "death tax."[40] And while a number of right-wing blogs provide you with many reasons for ending the estate tax, these tend to be beholden to the rich. The Republicans clamor for an end to the inheritance tax altogether—that would produce by far the greatest benefits for the rich. But they don't have a clue how to make up for the inevitable budget deficit that would result. However, Vice President Dick Cheney voiced the Republican view when he said: "deficits don't matter." The Democrats tend to resist any reduction in the inheritance tax—which shows that they just don't understand the current increase in the number of estates of a few million dollars due to asset bubbles.

Table 1.4 Number of taxable estates by size in year 2000, prior to Bush tax cuts of 2001. The personal exemption was $675,000 (double for a couple) and the nominal tax rate was 55%

Size category ($ million)	Number of taxable estates	Average value ($)	Average tax paid ($)	Effective tax rate
0.6–1	18,634	847,947	41,270	4.9%
1–2.5	23,827	1,490,693	230,238	15.4%
2.5–5	5,917	3,424,938	858,768	25.1%
5–10	2,258	6,884,752	1,950,852	28.3%
10–20	814	13,553,285	3,608,721	26.6%
> 20	549	58,667,401	10,418.672	17.8%

Data on taxable estates for year 2000, prior to the Bush tax cuts of 2001 are shown in Table 1.4. Note that the larger estates paid a tax rate far lower than the official rate of 55% through use of charitable trusts and other tax-saving devices.

In 2001, the new estate tax laws pushed through the Republican Congress by President Bush provided the changes shown in Table 1.5. There have been wacky tax laws enacted as far back as history goes, but this one is clearly near the top for being weird, inconsistent and inexplicable. The 2001 tax change provides for several gradual increases in the estate tax exemption together with some moderate reductions in the tax rate. The exemption rises to $3,500,000 in 2009, and then the entire inheritance tax is eliminated in year 2010. However, in 2011, the deductible reverts to $1,000,000.

Table 1.5 Schedule for estate taxes as changed by the Bush 2001 tax cuts

Calendar year	Estate exemption	Highest estate tax rate (%)
2001	$675,000	55
2002	$1,000,000	50
2003	$1,000,000	49
2004	$1,500,000	48
2005	$1,500,000	47
2006	$2,000,000	46
2007	$2,000,000	45
2008	$2,000,000	45
2009	$3,500,000	45
2010	N/A no estate tax	
2011	$1,000,000	55

Table 1.6 Effect of raising estate tax exemption

Year	Personal exemption (double per couple)	Percent of estates subject to estate tax	Number of estates subject to estate tax
2000	$675,000	2.2	52,000
2003	$1,000,000	1.6	37,100
2006	$2,000,000	0.5	12,600
2009	$3,500,000	0.3	7,100

Thus, if you have a sizable estate and you are in ill health, you owe it your heirs to commit suicide in 2010 in order to double their inheritance as compared to dying in *2011*.

With the bubbles in asset values for stocks and real estate over the past few decades, the number of estates exceeding $1,000,000 has increased remarkably. Indeed, in my hometown of South Pasadena, CA, the average price of a house is close to $1,000,000. Almost anyone who owns a house and has been living in it for some time is a "millionaire." In fact, the term "millionaire" is no longer exclusive as it was in the past. Raising the estate tax exemption would allow estates of a few million dollars to escape the estate tax, while preserving the bulk of federal revenues from large estates. The effect of raising the estate tax exemption is shown in Table 1.6. By raising the personal exemption from $675,000 to $3,500,000, more than 85% of previously taxed estates would be eliminated from paying estate taxes, while the largest estates would remain subject to the estate tax (see Tables 1.7 and 1.8).

Taxable farm and small business estates have been a source of concern because it is claimed that estate taxes might cause them to have to liquidate these businesses. According to the Center on Budget Policy Priorities (CBPP):

Had the 2006 exemption level of $2,000,000 ($4,000,000 per couple) been in place in 2000, the number of taxable farm estates would have dropped by more than 90 percent, and the number of taxable family-owned businesses by almost 75%. At an exemption level of $3,500,000 ($7,000,000 per couple), as will exist

Table 1.7 Dependence of number of estates subject to estate tax on exemption level

Year	Personal exemption (double per couple)	Estates less than $5,000,000	Estates more than $5,000,000
2003	$1,000,000	31,900	7,700
2006	$2,000,000	5,100	4,900

Table 1.8 Effect of raising the estate tax exemption on farm and small business estates

Year	Personal exemption (double per couple)	Number of farm estates subject to estate tax	Number of small business estates subject to estate tax
2000	$675,000	1660	485
2006	$2,000,000	125	135
2009	$3,500,000	65	95

in 2009, fewer than 100 family businesses and only 65 farm estates would have paid any estate tax. The estate tax changes made so far have been well targeted, providing the bulk of the relief to smaller estates and preserving a large share of estate tax revenue. The changes in the estate tax that have taken place since 2001 have exempted many estates from tax and provided tax reductions to other estates that remain taxable. In 2006, nearly four-fifths of the benefits of these changes will go to estates valued at less than $5 million. Further, because the changes made so far focus on raising the exemption level rather than sharply reducing the tax rate, permanent reform along these lines would preserve a large share of estate tax revenue.[41]

As the CBPP showed with ample data, increasing the exemption benefits the smaller estates, while decreasing the estate tax rate overwhelmingly benefits the larger estates. That is why Republicans (like Senator Jon Kyl) who represent the rich, advocate reducing the estate tax rate. The differences between permanently adopting the 2009 estate tax parameters vs. total elimination of the estate tax are:

- Use of the 2009 estate tax would preserve 60% of the revenues lost by total elimination.

- 96% of benefits of total elimination of the estate tax would accrue to the largest estates.

THE ALTERNATIVE MINIMUM TAX

The AMT grew out of a minimum tax that was first enacted in 1969 to ensure that the highest-income households could not exploit loopholes, exclusions, and deductions to avoid paying any federal income tax. It has been claimed that this legislation was devised by the Democrats to prevent 155 extremely wealthy families from avoiding income taxes. However, as is usual with tax legislation, the rules were not planned well, and they failed to index the AMT for inflation. As a result, although it initially applied to relatively few taxpayers, the number of taxpayers who owe extra taxes due to the AMT grew over the years, reaching 4,000,000 in 2006, and would have grown to 25,000,000 in 2007 were it not for a one-year "band-aid" fix enacted at the end of 2007. This situation was exacerbated by the Bush tax cuts of 2001 that reduced ordinary income taxes but failed to modify the AMT. Because about 2/3 of AMT taxes derive from elimination of state and local taxes, states with high state income taxes have the greatest number of people subject to the AMT. These states (California, New York, New Jersey, Connecticut) tend to be Democratic states and the Republican administration of 2000–2008 was not highly motivated to change the AMT. President Bush's 2004 State of the Union address did not even mention the AMT.

As the years went by, the AMT began to impact more and more taxpayers in the middle class, causing outcries of resentment. Under the AMT a taxpayer calculates his income taxes with and without the AMT and pays the higher of the two. The AMT eliminates some deductions, exemptions and credits, such as the deduction for state and local taxes and the personal exemptions. It has been estimated that in 2005, people subject to AMT paid an average of $4,350 on top of their regular tax. In 2006, almost 12 percent of married couples with children earning $100,000 to $200,000 owed AMT. Before 2000, the AMT never affected more than 1 percent of taxpayers.

The AMT provides a fixed exemption, and only considers income above that amount. The exemption for married couples filing jointly was originally set at $40,000 in 1987 but it was not automatically indexed for inflation. It was held constant at $40,000 through 1992, was upped to $45,000 from 1993 to 2000, and then was tweaked in a series of steps after 2001. The exemption amount was increased to $49,000 for 2001, $58,000 for 2003 through 2005, and to $62,550 for 2006. However, for reasons that are difficult to fathom,[42] the exemption for 2007 was scheduled to revert all the way back to the 2000 level: $45,000. That would have made an additional 20,000,000 taxpayers susceptible to the AMT. Had the government allowed this to happen, there would likely have been a taxpayer rebellion. The Congress was unable to deal with this impending political disaster for almost all of 2007. Finally, with great fanfare, at the eleventh hour in December of 2007, the Congress passed a one-year band-aid patch to the AMT for 2007, raising the exemption for married couples in 2007 to $66,250, thus proclaiming a great victory. However, roughly 4,000,000 people still paid the AMT in 2007, and basically it was little different from 2006. One element that held up passage of the band-aid to the AMT was that the Democrats wanted to raise taxes on the rich to compensate for the loss of federal revenues due to raising the AMT exemption, while the Republicans opposed this under the slogan "No new taxes." Besides, as Cheney said: "Deficits don't matter." Finally, both parties, under pressure from the populace, reluctantly agreed to band-aid the AMT for one year. In 2008, the exemption was raised slightly to $69,970. The probable outcome will be that the Democrats will raise the AMT exemption permanently, and raise taxes on the rich to compensate. The Republicans will adopt the mantra that "Democrats raised taxes" and repeat this a million times, thus winning Congress in the 2010 election.

The Bush tax cuts of 2001 and 2003 cut ordinary income taxes but did nothing about the AMT. As a result, the nominal tax cuts for millions of taxpayers were more or less erased because they had to pay the AMT. To soften this blow, Congress passed some temporary increases in the exemption used to calculate the AMT, but the number of taxpayers paying the AMT still doubled from 2,000,000 to 4,000,000 from 2002 to 2006. As we pointed out previously, one of the oddities of the 2001 and 2003 tax cuts was the fact that they expire at the end of 2010. It appears that the reason for this was the propaganda value of limiting the cumulative deficit predicted to result from the cuts. But the problem here is that the government has become so dependent on the AMT as a source of revenue that it may have difficulty reducing the impact of the AMT on middle class wage earners, without raising taxes on the rich—which seems unlikely. If the Bush tax cuts are extended beyond 2010, and the AMT is not overhauled, the number of taxpayers who pay the AMT could increase to about 40,000,000 in 2012. The cost to the government of reducing the number of taxpayers subject to the AMT while at the same time retaining the Bush tax cuts after 2010, would be hundreds of billions of dollars per year.

The deficits produced by the Bush tax cuts of 2001 and 2003 were made palatable in the short run by (a) excess SS collections, and (b) expansion of the AMT, but they still ran up substantial deficits. The Bush tax cuts are clearly untenable for the intermediate run, and amount to a sort of Ponzi scheme for temporary tax relief for the rich while dumping on the poor and the middle class.

INCOME TAX BRACKETS AND BUDGET DEFICITS

In the 1920s there were as many as 56 income tax brackets divided by $2,000 increments in earnings. Over the years, the number of brackets was gradually reduced. Today, we have only four brackets. Except for a three-year period beginning in 1988, the income tax structure always provided that the tax rate was higher in each successively higher income bracket. However from 1988 to 1990, there was a "bubble" in which medium incomes paid a higher rate than higher incomes.

The history of variation of the tax rate in the highest income bracket is shown in Figure 1.4. From 1917 through 1986, a period of 70 years, the maximum income tax rate in the highest bracket was 50% or higher, except for the interval 1925–1931 when it was 25%. The low income tax rate in the late 1920s appears to have contributed to the speculative excess that led to the crash in 1929. It also helped create an inordinate number of millionaires (in 1920s dollars) and increased the divergence between the rich and the poor.

The top tax rate was over 90% from 1944 to 1963, and was 70% or higher from 1936 through 1970. The cultural norm for much of the 20th century was for high maximum tax rates on the higher income brackets. The lowest income tax rates since 1917 prevailed from 1988 to 1992. President George Bush (senior) was roundly vilified for the increase in the maximum tax rate from 28% to 31% in 1991 and he later issued many a mea culpa for this "mistake." Maximum income

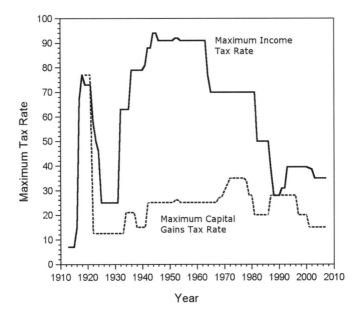

Figure 1.4 Income tax rate (%) in the highest income bracket for each year since 1913. Also shown is the maximum long-term capital gains tax rate.

tax rates since 1988 have ranged from 30% to 40%, about half of what they were earlier in the 20th century.

It appears likely that the low maximum tax rates prevailing in the late 1920s and the late 20th century contributed to the formation and expansion of stock market bubbles.

The data on annual increases in federal debt are plotted in Figure 1.5. It can be readily seen that the large deficits began in the 1980s when the maximum federal income tax rate was reduced. The dot.com stock market bubble artificially expanded government revenues during the late 1990s. According to Hodges,[43] during the four-year period FY 1998–2001, politicians claimed a $557 billion surplus, yet total debt increased $438 billion in that period—meaning that actual debt reduction was nearly $1 trillion over-stated. In general, there is a considerable anti-correlation of federal budget deficits with maximum income tax rate. On January 7, 2009 President-Elect Obama predicted that there would be "annual trillion-dollar deficits for years to come."

The transition to lower maximum income tax rates coincided with the rise to power of Republicans (Reagan 1981–1988, Bush (senior) 1989–1992, Bush (junior) 2001–2008). While Bill Clinton, a "centrist" Democrat, was in power from 1993–2000, the maximum tax rate underwent a moderate increase to 39.6%.

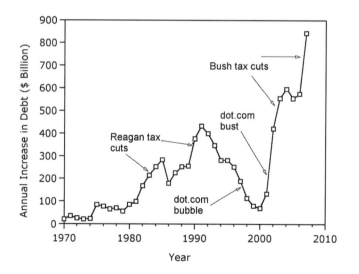

Figure 1.5 Annual increases in federal debt. The dot.com stock market bubble artificially expanded government revenues during the late 1990s.

The justifications for lowering the maximum income tax rates during last decades of the 20th century, were based – at least partly – on the Republicans' claim that federal government tax receipts would actually increase because the lower tax rate would stimulate investment, leading to economic prosperity and higher tax revenues despite the lower tax rate. This claim was made in conjunction with the "trickle-down theory" that held that if we could only succeed in making the rich, rich enough, they would invest their excess funds in enterprises that would create jobs and trickle wealth down to the middle class and the poor. Actually, the policies since 1985 have produced bubbles and a trickle-up result, making the rich richer than ever.

Aside from its effect on federal budget deficits, the lowering of maximum income tax rates (late 1920s and post-1985) benefited the rich by a large margin, produced the greatest number of super-rich people, and greatly exacerbated the disparity between the rich and the poor. The "trickle-down" theory has been shown to be a myth. Figure 1.6 shows the after-tax income of Americans since 1979 (in constant 2005 dollars). Incomes are divided into five quintiles, as well as the top 10%, 5% and 1%. In the lower brackets, incomes have hardly changed, whereas in the top 1% they have tripled. The incomes of the lower four quintiles are buried at the bottom of this graph. Note the large drop in income for the top 1% in 2001–2002. This occurred because they derive a high proportion of their income from stocks and the dot.com crash reduced profits in those years. Figure 1.7 shows the after-tax income of the lower four quintiles (in constant 2005 dollars). The gains in the lower quintiles have been modest, and the increases from 1979 to 2005 go up with rising quintile. During the post-1979 period, increases in real wages of lower-income and middle-income Americans were small as raises barely exceeded inflation. More families required the husband and wife to work to make ends meet, and working hours expanded for professionals. According to David Cay Johnston (DCJ), the average family today works 20 more weeks of paid labor than in 1975, and the wages and salaries of 99% of Americans stagnated or declined from 1973 to 1997. To the degree that there was prosperity, it was based mainly on asset bubbles in stocks and real estate, rather than wages.

CAPITAL GAINS

Historically, the capital gains tax rate has almost always been far lower than the ordinary income tax rate, as shown in Figure 1.4. Most recently, President Bush

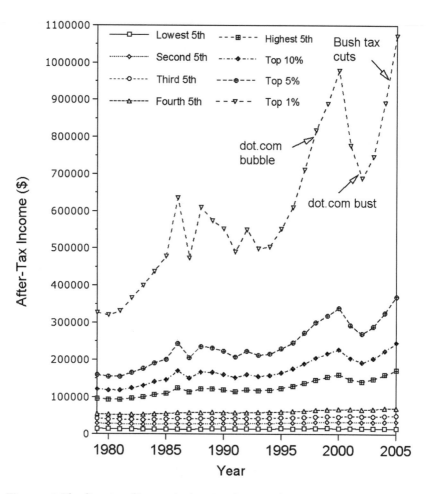

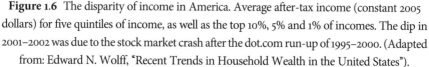

Figure 1.6 The disparity of income in America. Average after-tax income (constant 2005 dollars) for five quintiles of income, as well as the top 10%, 5% and 1% of incomes. The dip in 2001–2002 was due to the stock market crash after the dot.com run-up of 1995–2000. (Adapted from: Edward N. Wolff, "Recent Trends in Household Wealth in the United States").

triumphantly pushed through a reduction of the capital gains tax rate from 20% to 15% in 2001 based on the claim that it would promote business, whereas all it did was aid and abet bubble formation.

The Internet is full of learned articles by economists, typically in the employ of right-wing *think-tanks* that "prove" that lower capital gains taxes stimulate the economy and produce prosperity for all. There are even articles that purport to show that the 12.5% rate prevailing in the late 1920s did not contribute to rampant

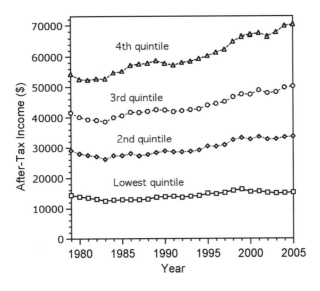

Figure 1.7 Average after-tax income (constant 2005 dollars) for the lower four quintiles of income. (Adapted from: Edward N. Wolff, "Recent Trends in Household Wealth in the United States").

speculation. Recently, the CBPP[44] discussed the economic impact of lowered capital gains taxes. After a capital gains rate cut, there are two effects. One is that investors, fearing that the cut may be temporary, may take gains and pay taxes on them. In addition, the lowered rate will tend to promote asset bubbles, leading to further capital gains. Hence, the short-term response to a capital gains tax reduction is likely to be a temporary increase in revenues. However, in the longer run, it will decrease revenues.

But as Edward N. Wolff (ENW) has shown,[45] the rich in America own most of the assets susceptible to capital gains. Table 1.9 shows the distribution of stock ownership by wealth class. The upper 5% of households owns 65% of stocks and the upper 10% owns 79% of all stocks. The bottom 80% of households owns 10% of the stocks.

According to ENW, in 2004, the top 10 percent of households owned 80 percent of non-home real estate and 90 percent of the total value of stock shares, bonds, trusts, and business equity. The Los Angeles Times (1/10/09) reported that the top 0,190 of incomes accounted for 5090 of capital gains earnings, and the top 590 accounted for 9090.

While many economists argue that a low capital gains tax rate encourages business investment and enhances prosperity, in an era dominated by booms and bubbles, where investment in stocks and real estate are not made primarily for business expansion, but rather for speculation to turn over to a new buyer on the

Table 1.9 Percent of stock owned by various wealth classes

Wealth class	Percent of stock owned	Cumulative percent of stock owned
Upper 1%	37	37
Next 4%	29	65
Next 5%	14	79
Next 10%	12	91
Next 20%	7	98
Next 20%	2	99
Bottom 40%	1	100

manic upward boom, the main effect of low capital gains tax rates seems to be encouragement of speculation, feeding the bubbles and manias, and more profits for the rich.

SOCIAL SECURITY AND MEDICARE

In 1983, a study led by Arthur Greenspan concluded that the Social Security (SS) system would start running in the red in 31 years (2014) and that fundamental changes were needed to keep SS solvent. (Of course, Greenspan was mistaken, as he has been on most issues). Since then, the future demise of SS has been a recurring theme of Republicans. This seems strange in some ways because one might think that the Republicans would prefer a regressive tax on lower incomes that exempts the rich, although they may principally resent the employers' contributions, as well as Medicare—which has no earnings cap. As a result of the Greenspan study, the Democrats became concerned about the future of SS, and decided that the answer was to collect more than was paid out, and invest the surplus into a SS *trust fund* that would supposedly be available to cover future shortfalls. The excess of collections over payouts started modestly but built up substantially as the years went by. However the Democrats were not unanimous; Senator Moynihan called it "thievery." According to David Cay Johnston (DCJ), from 1984 to 2002, Americans paid in to SS $1.7 trillion more than they received in benefits. This supposedly went into the *trust fund*. Since then, the amount of money in the SS *trust fund* has been reappraised by a number of observers, including President George W. Bush, who mentioned the figure $2.6 trillion in a 2001 State of the Union address, although it appears that the proper value may have been only about $2 trillion at the end of 2006.

Various politicians have provided assurances to the public regarding the trust fund at various times. President Bush said in his 2001 State of the Union address:

> To make sure the retirement savings of America's seniors are not diverted in any other program, my budget protects all $2.6 trillion of the Social Security surplus for Social Security, and for Social Security alone.

However, there is a great deal of confusion about this *trust fund*, and Mr. Bush did not appear to understand it in 2001. The notion of a SS *trust fund* is not simple. Nevertheless, there are several websites that explain the matter fairly well.

The Social Security *trust fund* is not a trust fund in the usual sense. In an ordinary trust, the trust funds are separate and distinct, and are typically invested for the benefit of the trustees. However, the 1935 Social Security law requires that the SS trust funds be "invested in US Government securities." Because trust fund securities are themselves federal securities, they are essentially IOUs that the Government has made out to itself, and they do not necessarily increase its ability to pay benefits. These funds get lost in the federal government's coffers, and the only obligation to the SS system is the memory (easily forgotten) that in some distant unspecified future, the federal government owes these funds to SS. Some defenders of the present system point out that by diverting excess SS funds into the general coffers, they help reduce the federal deficit, making it easier in the future to repay the SS trust fund. But the only way that the federal government can repay those funds in the future is by raising taxes, and that does not seem to be politically viable.

A few years after President George W. Bush promised to put a "lock box" on the SS *trust fund*, he contradicted himself by saying:

> Every dime that goes in [to Social Security] from payroll taxes is spent. It's spent on retirees, and if there's excess, it's spent on government programs. The only thing that Social Security has is a pile of IOUs from one part of the government to the next.
>
> The money-payroll taxes going into Social Security are spent. They're spent on benefits and they're spent on government programs. ***There is no trust.***

That "there is no trust" in Mr. Bush is widely agreed upon. On the other hand, the SS *trust fund* seems to be included in the federal debt.

SS taxes are divided into two parts. The contribution to Medicare is 1.45% and is applicable to all wages without a cap. The regular SS tax is 6.2% of wages but is only charged up to a cap that was $97,500 in 2007 and increased to $102,000 in 2008. Wages above the cap do not pay the 6.2%. Thus the total of SS tax and Medicare tax is 7.65% of wages up to the cap, and this is matched by an employer's contribution.

In essence, the excess collections for SS are just another tax on the people to provide funds to run the government. But this is a very retrogressive tax, for it begins on the first dollar of earnings, and remains constant up to a fixed cap, while the rich pay no SS tax on almost all of their income (i.e. the portion of their wages greater than the cap, plus 100% of capital gains, dividends and other income). The SS tax is an income redistribution plan in which the poor receive more than they put in. The rich are protected because no capital gains, dividends or other income, or wages above the cap are subject to the SS tax.[46] The percentage of income paid to SS by a person with a high income is small. The cost of SS is born mainly by the middle class because they pay 7.65% up to $102,000 of earnings.

When SS and Medicare charges are added to the regular income tax brackets, the result is as shown in Table 1.10. The effect of adding SS and Medicare charges to the income tax rate (in 2007) is that incomes in the range $63,700 to $97,000 pay a higher total rate than incomes from $97,000 to $195,800. Overall, the effect of SS tends to equalize tax rates over $63,700. More importantly, the low taxes on capital gains and dividends amplify the differential between rich and poor.

> Since the excess collection of SS funds coincided in time with large tax reductions for the rich, and the net effect of the excess SS collections was to reduce the deficit produced by the tax reductions for the rich, we must conclude that the excess SS collections acted as an anti-Robin Hood scheme to rob from the poor and pay the rich. According to DCJ, the bite from SS is greater than that from income tax for 75% of Americans.[47]

Table 1.10 Comparison of total tax brackets when SS and Medicare are added to income tax

2007 income greater than:	Income tax rate (%)	Income tax + Social Security + Medicare (%)
$0	10	17.65
$15,650	15	22.65
$63,700	25	32.65
$97,000	25	26.45
$128,500	28	29.45
$195,850	33	34.45
$349,700	35	36.45

Robert Shiller[48] suggested that the design of SS should be improved. He began by affirming what SS should not do: it should not invest the SS trust funds in the stock market, as that would inject excessive risk and uncertainty into the system. However, he treated the SS trust funds as if they were real and tangible and it seems doubtful that such funds actually exist, except as paper notations on spreadsheets. If the SS System elected to invest, say, $1 trillion of its trust funds into the stock market, where would that money come from?

Shiller suggested that the SS tax rate and the payout rate could be made flexible to vary with economic conditions. He suggested that "contribution rates and benefit rates should vary over time depending on the relative needs of workers and retirees," depending on the consumer price index and per capita national income.

He seemed to be saying that contribution and benefit rates should be flexible, and should be periodically adjusted to keep the system solvent. This seems a bit naive. Projections for the future are unidirectional; the population is getting older. The cost of operating SS is going up. The only way to keep a balance for the future is to increase contributions, preferably by raising the cap on income and by taxing all income, not just wages.

Republicans hate SS. As we stated previously, this seems strange because one might think that the Republicans would prefer a regressive tax that primarily taxes middle-class incomes and exempts the rich. Republicans provide a continual barrage of warnings that SS is going broke and Democrats have wrung their hands and waffled in response. Most of these projections assume minimal increases in the SS tax. However, as the population ages, there seems little doubt that SS taxes will have to increase. What is rarely mentioned is that if the SS tax were applied to all income (not just wages under a cap) it would easily provide for future needs.

INEQUALITY

WHY INEQUALITY PERSISTS AND EXPANDS

JKG[49] commented at considerable length on inequality between the rich and the poor. He pointed out that over the years, few topics have generated more controversy than the proposition that "the rich should by one device or another share their wealth with those who are not." With tongue in cheek, he mentioned rather laconically that "the rich are opposed" to such a proposition for many and varied reasons, which ultimately boil down to their (natural) unwillingness to give up their advantages. In the same rather dry tone, JKG mentioned blandly that the poor favor greater equality.

A great source of consternation and puzzlement has always been the question: Since there are many more poor than rich, why don't the poor just tax the rich heavily and reduce the inequality? One can imagine several contributing factors as to why this has not happened, but none of these alone or taken together provide a satisfactory explanation. These factors are (in no particular order):

(a) The poor may be politically disorganized and are unable to marshal their forces into a concerted effort to soak the rich.

(b) The media are controlled by the rich, and they propagandize the poor, browbeating them into believing that the rich deserve to be rich.

(c) The various political legislatures and executives have mostly above-average income and assets and are beholden to the rich. They are prone to pass legislation favorable to the rich.

(d) The poor are afraid that if they embarked on a "soak the rich" campaign they would suffer persecution of one form or another.

(e) The Supreme Court, being subservient to the rich, would nullify it anyway.

(f) The poor are in awe of the rich and admire them too much to tax them.

(g) The rise and expansion of the middle class has diluted the influence of the poor. Members of the middle class may harbor hopes that they or their children might one day become rich.

To this list, JKG's point must be added:

In the US the poor have reacted sympathetically to the cries of pain of the rich over their taxes.

There are many defenders of inequality, historically, as well as in the present.[50] JKG provided a summary of their arguments in defense of inequality:

(a) There is a deep-rooted cultural belief that what a man lawfully earns or receives is rightfully his. This is regarded as an inalienable right akin to "life, liberty and the pursuit of happiness."

(b) Tampering with the system that allows the rich to hold on to their assets would break down the economic system, lead to chaos, and make things worse for everyone.

(c) Allowing the rich to remain rich provides an incentive for all people to strive to get ahead.

(d) The rich provide support for education and the arts.

(e) It would be very monotonous and boring if we all had the same income.

(f) The rich are needed for capital formation and investments. Having large concentrated blocks of money allows efficient investment.

(g) Equality raises the specter of communism and atheism.

According to JKG, the liberal attitude toward inequality has consisted mainly of uttering platitudes with only minor action. JKG summarized the liberal view: "It is terribly uncouth to soak the rich." Meanwhile, as JKG pointed out, the rich continue to get richer.

The rich have out-maneuvered the liberals using clever political tactics. The religious conservatives in America, as typified by the Southern Baptists and a wide swath of Protestants in the Midwest place social issues such as anti-abortion and anti-gay rights at the top of their agendas. Their underlying motivation seems to be an antagonism toward the sexual freedom that has developed in our culture over the past several decades, and their antagonism to abortion seems to be vested more in the hope of making sex more problematic than in preserving lives.[51] Nevertheless, regardless of their motivations, they have been co-opted by the Republican Party, which primarily represents the rich, and together have created a formidable political machine composed of strange bedfellows (not literally). Strange indeed, considering that many of the Southern Baptists have modest incomes, yet they have aligned with the party of the rich because it panders to their reactionary social attitudes.

In the 19th century and the first half of the 20th century, the topic of inequality was discussed widely. Ending or reducing inequality was a prime motivating factor in the emergence of communism and socialism. As JKG pointed out, "the decline of interest in inequality cannot be explained by the triumph of equality" for the divergence between the rich and the poor in America is higher today than it was even in the roaring twenties, and the trend seems to be toward increasing the gap between rich and poor. Much of this divergence is due to the changes in the income tax structure that have been enacted primarily by Republicans but often with complicity by Democrats.

JKG offered several reasons why inequality has faded out as an issue:

(a) Inequality has not produced the kinds of violent reactions predicted in the past.[52]

(b) As time went by we found that envy tends to be localized; perhaps we tend to envy our neighbor's new luxury vehicle, but not the wealth of an unseen billionaire in some remote location.

(c) A more credible point made by JKG is that in the past, the rich were directly involved in the corporations that employed the poor and the middle class. Today, the rich are mostly separated from the corporations, and the class struggle, to the small degree that it remains, is between the poor and the corporations, not between the poor and the rich.

The fading of inequality as an issue may also be tied to the rise of a large middle class, and the general and widespread disillusionment with communism and socialism in the world.

JKG concluded that that the ancient "question of whether the rich are too rich remains irresolvable."

Delamaide said:[53]

They [Bankers] believe that man should be free to earn, to save, and to invest his money. Self-interest motivates the market place that magically allocates resources in the most efficient manner. Wealth is created. The rich get richer, but so do the poor, which does not exclude their getting relatively poorer.

INEQUALITY OF WEALTH IN THE UNITED STATES

Edward N. Wolff[54] (ENW) has written a number of books, reports and papers on wealth in households in the United States. He pointed out that whereas most studies that dealt with the distribution of "well-being" concentrated on the distribution of income, there are a number of reasons why family wealth (assets) may be a better indicator of *well-being*. As ENW showed, "the only segment of the population that experienced large gains in wealth since 1983 is the richest 20 percent of households."

ENW used the term *wealth* in the limited sense as marketable wealth (or net worth), defined as the value of all marketable assets less debts. Net worth is thus [assets—liabilities]. Total assets include: (1) owner-occupied housing; (2) other real estate; (3) cash and bank deposits; (4) bonds and other financial securities; (5) cash surrender value of life insurance and pension plans; (6) corporate stock and mutual funds; (7) equity in businesses; and (8) equity in trust funds. Liabilities include: (a) mortgage debt; (b) consumer debt; and (c) other debt.

ENW did not include consumer durables such as automobiles, televisions and furniture, in wealth, because these items are not easily marketed. As he said: "their resale value typically far understates the value of their consumption services to the household." But that seems illogical because it is not *wealth* so much as *well-being* that we are interested in, and having such assets improves one's well-being even if it is not marketable. If these assets had to be replaced, the replacement cost would severely impact one's *well-being*. ENW also excluded the value of future social security and retirement benefits from private pension plans because they cannot be marketed. But such funds significantly improve one's *well-being*, even if they are not immediately accessible.

ENW also used a concept of *non-home wealth* that omits the value of the domicile in which a family lives because a residence is not very liquid, and besides, one has to live somewhere. However, owning a more valuable house typically adds to one's *well-being*.

In comparing income and wealth over many years, one typically puts all years on a comparable basis by correcting for inflation with a consumer price index deflator. ENW discussed pros and cons of different indices; he used the standard CPI-U deflator.

ENW estimated the distribution of wealth and income in the United States for 2004 as shown in Table 1.11. These figures show that net worth was concentrated in the rich, with the top 1% owning 34% of national net worth and the top 10% owning 71% of national net worth. The bottom 40% had essentially no net worth. The figures are even more extreme for non-home net worth because poorer people tend to have most of their assets invested in their residences. The distribution of income is not so extreme, but still peaked near the top. Evidently, the rich manage to own almost everything but do not pay taxes on a proportionate share of income.

The changes in net wealth from 1983 to 2004 were concentrated toward the top of the distribution. Table 1.12 provides data on the distribution of the gain in net worth

Table 1.11 Distribution of wealth and income during 2004 (ENW)

	Percent of national net worth	Percent of national non-home net worth	Percent of total income
Upper 1%	34	42	20
Next 4%	25	27	15
Next 5%	12	12	10
Next 10%	13	12	14
Top 20%	85	93	59
Next 20%	11	7	19
Next 20%	4	1	12
Bottom 40%	0	−1	10

from 1983 to 2004. The top 10% of households accounted for 76% of the gain in total wealth, 81% of the gain in non-residential wealth, and 68% of the gain in income. The bottom 60% had essentially no gain in wealth and a very minor gain in income.

The number of American billionaires was relatively constant at around 13 from 1982 to 1985. After 1985, one estimate indicates that the number of billionaires increased almost linearly by about 17 per year, bringing the total to about 100 billionaires in 1990. Various estimates suggest that the number of American billionaires had grown to between 300 and 400 in 2007.

During the period 1983–2004, the number of households with net worth exceeding one, five or ten million dollars increased substantially, as shown in Table 1.13. Most of this was due to inflated stock and real estate assets.

Table 1.14 provides data on how wealth is distributed amongst various assets for households in different wealth groups. In the broad middle class (21st to 80th percentiles) the principal residence accounts for about 2/3 of total assets, and the

Table 1.12 Increase in wealth and income from 1983 to 2004 (ENW)

	Percent of total gain in net worth	Percent of total gain in non-home net worth	Percent of total gain in income
Upper 1%	35	41	33
Next 4%	28	28	22
Next 5%	13	12	13
Next 10%	14	12	13
Top 20%	89	93	81
Next 20%	10	7	13
Next 20%	2	1	4
Bottom 40%	−1	−1	2

Table 1.13 Number of households with net worth exceeding one, five or ten million dollars from 1983 to 2004 (ENW)

	Thousands of households	The number of households (in 1,000s) with households net worth equal to or exceeding (in 1995$)		
		$1 Million	$5 Million	$ 10 Million
1983	83,893	2,411	247	67
1989	93,009	3,024	297	65
1992	95,462	3,104	277	42
1995	99,101	3,015	474	190
1998	102,547	4,783	756	239
2001	106,494	5,892	1,068	338
2004	112,107	6,466	1,120	345
% Change	33.6	168	353	419

sum of residence, liquid assets and pension accounts amounts to over 86% of assets, while stocks and mutual funds are 4.2%. By contrast, the wealthiest 1% of households has much of its assets (76%) concentrated into unincorporated business equity and corporate stocks and mutual funds.

Table 1.14 Elements of household wealth (2004) (ENW)

	Percent of total assets			
	All households	Top 1% of households	Next 19% of households	21st to 80th % of households
Principal residence	33.5	10.9	32.2	66.1
Liquid assets	7.3	5.1	8.6	8.5
Pension accounts	11.8	5.3	16.0	12.0
Corporate stock and mutual funds	17.0	26.9	16.3	4.2
Unincorporated business equity	28.6	49.3	25.4	7.9
Miscellaneous assets	1.8	2.6	1.5	1.4
Total Assets for Each Category	100	100	100	100

DEBT

A number of studies, reports and websites have raised alarms about the levels of debt in America in the late 20th century and early 21st century.

Debt may have many virtues in the right circumstances. For example, it makes complete sense that families should borrow to purchase a primary residence, taking up to 30 years to repay the loan, because (a) it is reasonable to expect continued employment and earnings for such a period, (b) in any event, the value of the loan is protected by the inherent value of the residence,[55] and (c) real estate values tend to be relatively stable.[56] It would not make any sense at all to require that residences be purchased for cash; very few would be sold. Similarly, a corporation that uses debt to invest in new facilities, capabilities or ventures with the reasonable prospect of recouping much more than the amount borrowed, would be using debt effectively. However, when debt is used to fund operating expenses (as opposed to capital investments with prospects for increased future payoff) then there is typically little prospect for repaying the debt, and it amounts to a swindle in the sense of "Speculations, Bootstraps and Swindles". When governments borrow in order to build infrastructure (whether it be a bridge, a water system, or whatnot) with the intent to pay it back from credible tax levels, it is appropriate and sensible. However, when governments operate with a continuing deficit and borrow to cover operating expenses with little hope of repayment from future revenues, that amounts to a Ponzi scheme.

US FEDERAL DEBT

The history of US federal debt is shown in Figures 1.8 and 1.9. As Steve McGourty[57] pointed out, since 1946, Democratic Presidents increased the national debt at an average of 3.2% per year while Republican Presidents increased the national debt at an average of 9.7% per year while in office. Republican Presidents out-borrowed Democratic Presidents by a 3:1 ratio. Debt has been on a steady incline ever since the Reagan presidency. The only exception to the steep increase over the last 25 years was during the Clinton presidency, when the combination of revenues from the dot.com bubble, combined with excess SS collections, temporarily reduced the rate of increase of debt. The Republican administrations, particularly those of Mr. Reagan and Mr. Bush (Junior), advanced the federal debt the most. Although Republicans always accuse Democrats of following a "tax and spend" policy, Republicans follow a "spend and borrow" policy.

But these figures are given in current year dollars. If we corrected them for inflation, the curve would flatten out considerably. Dollars in 2007 are worth about 1/5 of 1970 dollars so the equivalent debt in 2007 would be about 1/5 of that given in the figures if the debt were expressed in 1970 dollars.

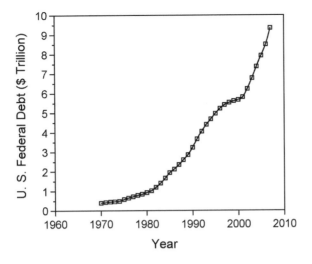

Figure 1.8 History of federal debt (linear scale). (Based on data from: http://mwhodges. home.att.net/nat-debt/debt-nat-a.htm).

Nevertheless, if we divide the debt of $10 trillion (at the end of 2008) by the population of 300 million we obtain the debt per person = $33,000.[58]

It is difficult to construct a detailed breakdown of the federal debt. The federal debt at the end of 2008 is claimed to be a bit over $10 trillion. Of this amount,

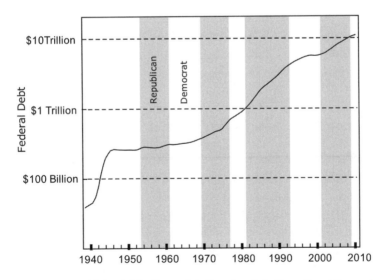

Figure 1.9 History of federal debt (log scale). Gray bands indicate Republican administrations. (Adapted from S. McGourty)[61].

approximately $6 trillion was conventional debt owed to the public in the form of treasury bonds and T-bills. Of that $6 trillion, nearly half, about $2.9 trillion, was owed to foreign interests. The annual interest on this debt is something of the order of $240 billion, for an average interest rate of about 4%. That part is easy to understand. The difficult part is the other $4 trillion of federal debt. Apparently,[59] this element of the federal debt includes the Social Security trust funds, which represent paper transactions between intergovernmental agencies.

According to one website:[60]

> The Social Security trust funds are United States Treasury bonds. These bonds are issued by the US Treasury to raise money to pay for budget deficits. The total value of all outstanding Treasury bonds is the national debt. The Social Security trust funds own part of the national debt.

A United States Treasury bond is a promissory note made out by the Government with an obligation to pay a sum to the holder of the bond. No such documents exist in the case of the putative government bonds "issued" to cover the excess payments into SS. And if they did exist, they would be payable by the Government to itself! Thus, roughly $2 trillion of this $4 trillion is accounted for by the Social Security trust funds. Does it make sense to include this in the national debt? The other $2 trillion seems to be in the form of loan guarantees for residential mortgages and public works, but it is difficult to pinpoint these exactly. They are described as: "intra-governmental holdings also include other federal entity holdings such as revolving funds, special funds and Federal Financing Bank securities, as well as other government accounts."

Federal debt is sometimes discussed in proportion to the gross domestic product (GDP). Currently, the federal debt is about 68% of the GDP. On that basis, some would argue that the federal debt is not excessive. The argument goes that if the total debt of an entity were 68% of the annual income of the entity, there would be every reason to believe that it could someday be paid back. This might be based on the analogy to debt for a household. It would not be considered excessive for many households to have total debt equal to 68% of annual income. Indeed, household mortgage debt (backed by real estate value) might be 5–6 times annual income, with a reasonable expectation to be paid back over several decades. However, household consumer debt (backed only by a promise to pay) of 68% of annual income would certainly be considered to be excessive. In the case of federal government debt, the comparison to the GDP may be irrelevant if the government is unable to pay back its debt due to the political impediment to raising taxes to increase revenues. Furthermore, the GDP is not the Government's revenue. The Government's revenue is the tax it applies to the GDP.

As we discussed in "Speculations, Bootstraps and Swindles," K&A distinguished between three levels of speculation. In my terminology these are (1) speculations, (2) bootstraps, and (3) swindles (usually Ponzi schemes). These are distinguished by the likelihood of earnings from the venture being sufficient to pay back principal and interest on loans made to finance the venture. In a speculation, there is a reasonable prospect that if all goes well, the operating income from the enterprise will be sufficient to pay off both the interest and amortization of its indebtedness. In a bootstrap operation, it is likely that anticipated operating income will be sufficient so it can pay the interest on its indebtedness. However even making favorable assumptions, it is unlikely that the operating income will cover the amounts of principal due on maturing loans. In a swindle, the anticipated operating income is not likely to be pay the interest and principal on its indebtedness on the scheduled due dates; to get the cash the firm must continually increase its indebtedness until lenders will no longer support the venture.

The US federal debt must be classified as a swindle because there is no credible scenario by which the Federal Government can repay its debt without raising taxes inordinately, which would be political suicide and bring on an economic depression. Hence, the Federal Government can only repay presently maturing debt by further borrowing, thus borrowing from Peter's children to pay Peter. That is a classical Ponzi scheme. Fortunately for the US Federal Government, there is enough tension in the world that many foreigners view US Government securities as a "safe haven." Otherwise the dollar would fall even further.

One may now ask whether through good management and efficient operation, the Government can actually operate at a budget surplus, as has been claimed by some during the Clinton years. However, as Figure 1.5 shows, deficits were reduced from 1997 to 2000 but not eliminated. These reductions were due primarily to temporarily very high revenues from dot.com profits (i.e. just another bubble). Aside from the fact that dot.com revenues were incredibly high those years, excess social security collections (above payouts) amounted to a couple of hundred billion per year, and these were counted as "income" in claiming a drop in the deficit. But these assets are balanced by a liability that is owed to the SS Trust Fund, so claiming that it reduced the deficit is fakery. In actual fact, the reported national debt increased each year from 1997 to 2000, although it almost broke even in year

2000 when the NASDAQ Index rose 85% in one year to 5200. The bottom line is that were it not for a temporary dot.com bubble and the claim that the collection of SS Trust funds was government income, there would have been much greater deficits. There does not seem to be any sustainable way for the US Government to pay back its loans.

The Republicans would have you believe that by lowering tax rates, the economy will boom so much that despite the lower rates, government revenues will actually increase, producing a surplus of government funds. Data show that this has never happened and the national debt has actually increased more during Republican administrations than in Democratic administrations (see Figure 1.9, in "US Federal Debt"). The situation of the Federal Government is that it is caught between a rock and a hard place. It must borrow to operate. While theoretically, it might be able to pay back its loans (slowly over a period of time) by raising taxes, the political reality is that such a tax increase would be political suicide, and aside from the politics, would probably produce a severe depression, thus further exacerbating the difficulty of paying back loans. Hence we have a conundrum. The credibility of the Government to borrow depends on its implicit capability to tax, and thereby repay loans. However, the political and fiscal realities suggest that the Government can never pay back those loans. It must keep borrowing via new loans to pay back the old. That fits Minsky's definition of a Ponzi scheme. The Government must go into deeper and deeper debt forever. The real question is this. Given these facts, why do investors loan money to such a Ponzi scheme at such low interest rates? Perhaps the answer is that with a seemingly permanent state of tension, war, and uncertainty in world affairs, the US Treasury is viewed as the only safe haven for funds. Money is invested in US treasuries, not so much for profit, but principally for safety.

In discussing the national debt that in 1983 was a mere $1.39 trillion, Dela-maide:[62] said:

> A national debt of this proportion seems a fiction, that has no impact in the real world. But it does, and the Reagan administration adopted a fiscal policy that promised to add mind-boggling sums to unreal numbers. It was financial overkill, a mega-debt as difficult to think about as the unthinkable dimensions of nuclear arms.
>
> It is practically inconceivable that the federal government of the United States . . . would go broke. It was at one time practically inconceivable that a big, rich city like New York, the very financial capital of the world, could go broke, but it did.

Gerald Ford said:

> If we go on spending more than we have, providing more benefits and services than we can pay for, then a day of reckoning will come to Washington and the whole country just as it has to New York City.

Delamaide quoted Henry Kaufman, as saying: "The debt burden today is awesome and its constrictiveness is permeating our economic life" and suggested that "a depression from a deflating debt bubble might be beyond the power of the authorities to counteract it." At that time, US debt was about $1.3 trillion. In 2008, the US debt was more than $10 trillion. The candidates for presidency in 2008 had slightly different prescriptions for revitalizing the moribund US economy. However, both Obama and McCain shared one basic principal: they both would increase spending and thereby expand the national debt. Leonard Burman of the Tax Policy Center estimated that Obama's plan would add 3.4 trillion dollars of debt over the next decade and McCain's plan would add 5 trillion dollars of debt over that period.[63]

To add to our woes, a recent book suggests that the real cost of the Iraq War will be about $3 trillion.[64]

STATE AND MUNICIPAL DEBT

There has been a rapid increase in debt by American state governments since 1990. Table 1.15 provides data on US state government debt. Cumulative debt doubled from 1990 to 2004, and annual borrowed funds in 2004 were 3.5 times the level of 1990. Cumulative debt of the states totaled almost $2 trillion in 2004. In California, the predicted budget deficit for 2008 was about $16 billion (about $500 per capita) although it now appears that it might be closer to $20 billion. The response of the California State Government to this problem was to propose new expensive

Table 1.15 Annual borrowing and accumulated debt of the states (Statistical Abstract of United States (2008))

Item	1990	2000	2003	2004
Borrowed funds (total) ($ billion)	41	76	135	141
Debt at year end ($ billion)	858	1,451	1,812	1,951
Borrowed funds ($ per capita)	165	270	464	480
Debt at year end ($ per capita)	3,449	5,159	6,234	6,646

large-scale health programs that would add to the deficit. Other states have smaller deficits, but the CBPP estimated that the total budget shortfalls in 2008 for 25 states is about $41 billion dollars.

In addition to explicit debt, there is also an implied debt to cover future retirement and health care obligations, as discussed in "The Public Sector," where it is shown that for New Jersey, the retirement and health care obligations are greater than the actual cumulative debt.

Debt for the 35 largest counties in the United States totaled $70 billion in 2004.

Debt for the 35 largest cities in the United States totaled $220 billion in 2004. In New York City, the debt was $76 billion, or $9,000 per capita. In Los Angeles and Chicago, the debt was about $15 billion each, for a per capita amount of roughly $5,000.

HOUSEHOLD AND MORTGAGE DEBT

Household debt is divided into mortgage debt and consumer debt.

Consumer debt is divided into revolving debt (mainly credit card debt) and non-revolving debt (automobile loans, student loans, and other supposedly one-time purchases). Data on consumer debt are provided in Table 1.16. On a per capita basis, that amounts to over $8,000, and on a household basis it amounts to over $20,000.

Consumer debt tripled from 1990 to 2007, but in constant 1990 dollars, it increased by only 60% over that time period. While the level of consumer debt has been and remains high, the rate of increase does not seem onerous.

The interest on home mortgage debt is deductible from income tax, and therefore it is very beneficial for homeowners to convert other forms of debt to

Table 1.16 US consumer debt (billions of dollars)

Year	Revolving	Non-revolving	Total consumer
1990	251	554	801
1995	464	659	1123
2000	664	869	1533
2002	749	1225	1974
2003	771	1308	2079
2004	800	1391	2191
2005	825	1460	2285
2006	875	1512	2387
2007	922	1566	2488

mortgage debt. Over the past two decades, Americans have proceeded to do just that. However, when interest rates were dramatically lowered by the Fed in 2002 in a frantic effort to stem the collapse of the dot.com bubble, Americans went on a refinancing spree, and in the process, markedly increased their loan principals. This, coupled with the housing bubble that ensued, raised mortgage debt to unprecedented levels as shown in Table 1.17.

Under "normal" conditions, a modest rise in mortgage debt would not be alarming because the value of a loan is protected by the inherent value of the residence, and real estate values tend to be relatively stable. In "normal" times banks only loan up to ∼80% of the appraised value of a house, and appraised values tend to be conservative and stable. However, during the real estate bubble of 2002–2007 in the United States, loans were made for more than 100% of highly inflated appraised values, making these loans very risky. In the ensuing aftermath when the bubble popped in late 2007, real estate values went through significant declines.

Table 1.18 shows the annual net investment in mortgages. However, mortgage volume (sum of in and out) topped $2.5 trillion in 2006, the peak in the housing bubble.[66]

In the early 1990s, the annual investment in mortgages was around 5–6% of total mortgage debt and the net growth in mortgage debt was about 4% per

Table 1.17 Mortgage debt in America (billions of dollars)[65]

Year	Residential total	1–4 family	Multi-family
1990	2,903	2,615	288
1991	3,067	2,782	285
1992	3,219	2,947	272
1993	3,375	3,106	269
1994	3,553	3,283	270
1995	3,727	3,451	276
1996	3,963	3,675	288
1997	4,210	3,910	300
1998	4,600	4,266	333
1999	5,066	4,691	375
2000	5,514	5,110	404
2001	6,086	5,640	446
2002	6,856	6,371	485
2003	7,725	7,169	556
2004	8,847	8,238	609
2005	10,046	9,366	680
2006	10,921	10,190	731

Table 1.18 Annual investment in residential mortgages (billions of dollars)[67]

Year	Net mortgage investment
1990	236.0
1991	163.7
1992	152.7
1993	156.0
1994	177.5
1995	173.9
1996	236.0
1997	247.0
1998	389.9
1999	466.2
2000	448.3
2001	571.5
2002	770.1
2003	868.8
2004	1122.4
2005	1199.4
2006	874.9

annum. After the stock market crashed in 2001, the Federal Reserve System flooded the banks with low-interest funds and the Federal Government essentially stopped regulating banks in regard to their practices for making mortgage loans. From 2001 to 2007, annual growth in mortgage debt averaged over 12% of total mortgage debt. A significant fraction of these mortgage loans were swindles in the Minsky sense, because there was almost no hope that borrowers could repay the loans unless real estate continued to inflate at 10–20% per year.

Total household debt is the sum of consumer debt and mortgage debt. Household debt (including mortgage debt) over the past thirty years is summarized in Table 1.19. Household debt as a percentage of disposable income has crept up from 65% in the early 1980s to 127% in 2005. The average debt per household at the end of 2006 is estimated to be about $113,000 per household (including mortgage debt) based on the country's 114.4 million residences.

A number of websites have raised alarms regarding the levels of household debt in the United States. The publicity on the mortgage crisis of 2008 has focused on the large losses of major banking, mortgage and investment companies from the downward spiral in housing prices. However, the huge increase in household debt, particularly due to larger mortgages, has somehow been buried in this discussion.

Table 1.19 Outstanding household debt as percentage of disposable income[68]

	Household debt ($ billion)	Household disposable income ($ billion)	Debt as percent of disposable income
1965	330		
1970	455		
1975	736.3	1187.4	62.0
1980	1397.1	2009.0	69.5
1985	2272.5	3109.3	73.0
1990	3592.9	4285.8	83.8
1995	4858.1	5408.2	89.8
2000	6960.6	7194.0	96.8
2005	11496.6	9039.5	127.2
2006	13,000 (est.)	9600 (est.)	135 (est.)

Yet, the average household's equity in its residence dropped to less than 50% in early 2008, and is likely to drop a good deal further as real estate prices plummet. The additional mortgage debt acquired when house prices were rising remains as debt, but the equity backing up that debt is falling. Leverage plays a role. For example, if a California household owns a house that, at the top of the market in early 2007 was valued at say, $800,000 (a rather average house at 2007 prices) and the mortgage was $600,000, if the house price drops 25%, the equity in the house vanishes.

Thus we see that at all levels, federal, state, municipal and household, debt has increased remarkably since Mr. Reagan took office in 1980. Indeed, the hallmark of the Republican Party, though not admitted, is to borrow, borrow and borrow more. How much longer this can persist is difficult to predict. But as we stated earlier, when debt is used to invest in infrastructure with a reasonable prospect of repayment, the use of debt can be very constructive. However, when debt is used to fund operating expenses and there is little prospect for repaying the debt except by further borrowing, then it amounts to a swindle or a modified Ponzi scheme in the sense of "Speculations, Bootstraps and Swindles".

The Federal Reserve Bank of New York has held a rather optimistic view regarding the growth in household debt. Personal consumption expenditures make up about two thirds of the country's gross domestic product, and one of the great concerns of the Fed is question whether "the sharp increase of household indebtedness and the rising share of income going to payments on credit cards, auto loans, mortgages, and other household loans ... [produce] rising debt burdens [that would] precipitate a significant cutback in spending as apprehensive consumers take steps to stabilize their finances?"[69] This Fed document suggests that so long as the consumer spends, spends, and spends more, all is well regardless of the levels of debt. So, the concern of the Fed is whether the increase in debt load will affect consumer spending. This Fed document presents two alternative hypotheses:

(1) "Households may have taken on too much debt in recent years, placing themselves in a precarious financial position. Over time, these households will recognize that their indebtedness has made them more susceptible to financial distress in the event of a serious illness, job loss, or other misfortune. As a result, they will seek to reduce their vulnerability by paying down debt and decreasing their expenditures."

(2) "Households have willingly assumed greater debt in recent years because they expect their incomes to rise. They spend more in anticipation of increased earnings and they finance their higher spending through debt. Even if their incomes begin to fall, households may continue to increase their debt to maintain their spending—albeit at a reduced level—on the assumption that the income decline will be short lived. Only if the decline proves to be long lasting will households cut expenditures further and begin to pay down their debt."

What is most interesting about alternative (2) is that the "expected rise" can hardly be a rise in salary since real wages have been essentially flat for some time. The "expected rise" is more likely an expected rise in paper asset values (stocks and real estate), which in 1997 (when the Fed article was printed) was probably a widespread expectation with the stock markets booming. The Fed report concluded that their "analysis does not support the more alarmist view of debt" and that "overall, the model simulations suggest that there is little reason to expect that current debt burdens will trigger a decline in consumer spending." The scary part of this is that it is likely to be true. As consumers build up debt, they may be unlikely to cut back expenditures, and thus add further to this debt. From the Fed's point of view, there seems to be no limit to debt.

In a similar vein, the Federal Reserve Bank of New York has displayed a rather optimistic view of low savings rates.[70] This report claimed:

> The U.S. personal saving rate's negative turn in 2005 has raised concerns that Americans may have to curtail their spending and accept a lower standard of living as they pay off rising debts. However, a closer look at saving trends suggests that the risks to household well being are overstated. The surge in energy costs may have temporarily dampened saving, while the accounting of household income from stock holdings may be skewing saving estimates. Moreover, broad measures of saving have remained positive, and household wealth is on the rise.

After remaining around 10% from 1950 to 1985, the savings rate (as percentage of disposable income) decreased linearly after 1985 and finally went negative in 2005. One point made by the Fed is that capital gains from rising stocks made it unnecessary for many households to save in the conventional sense; however for people with lower incomes who don't own stocks (or don't own much), rising stock prices offered no cushion for the lowering of interest rates in savings institutions. While it may be true that "household wealth was on the rise," such paper gains mainly impact the wealthier segments of the population, and some of these gains were ephemeral, as we have subsequently seen in 2008. Nevertheless, the Fed seems to be happy with gains in paper assets replacing conventional savings. That such gains may be short-lived and not grounded in fundamentals seems not to be a concern of the Fed.

THE DOLLAR

For the past several decades, the United States spent a good deal more on imports each year than it collected from exports, resulting in a negative balance of payments. More than half of the trade deficit was due to importing oil. Foreigners end up holding large amounts of dollars, which they typically invest primarily in US Treasury securities, stocks and real estate. In a very uncertain and risky world, the US Treasury is a safe haven. Furthermore, the weak dollar provides a better yield (a European could acquire $1.58 for his Euro in early 2008, and invest these dollars in Treasury bonds at 3.7% to get an effective yield of 5.7%). For example, it has been estimated that at the end of 2007, China held $1.5 trillion in US currency. Is this a worrisome thing, or is it just business as usual?

The valuation of the dollar and the US trade deficit is a complex subject that defies easy explanation. As the Fed lowers short-term US interest rates, the value of the dollar to foreigners goes down. Since international oil sales are valued in dollars, the price of oil goes up. The US Federal Government continues to spend more than it takes in from taxes, and borrows the difference (about half from foreigners).

Michael Mussa discussed the question of the current account deficit at some length. According to Mussa, the US current account deficit was running at $900 billion per year at the end of 2006—equivalent to almost 7% of US GDP. Economists are divided as to the seriousness of this problem. Mussa described the two polar opposites of viewpoint as:

(1) Chicken Little (*the sky is falling*) vs. Alfred E. Newman (*what me worry?*)

The "sky is falling" advocates say:

> A day of reckoning is fast approaching when foreigners will no longer be willing to add rapidly to their already large net accumulations of US based assets. When this happens, the value of the dollar in foreign exchange markets will crash and the cost of capital to finance investment in the United States will shoot up, restricting demand within a much reduced supply of domestic and foreign saving. The US economy will fall into steep recession as investment and consumption spending are curtailed and the Federal Reserve is constrained by worries about the possible inflationary effects of a much weaker dollar. Meanwhile, a sharp fall-off in exports to the US will undermine growth in the rest of the world and threaten a serious global recession.

The "what me worry" advocates argue that there is no practical limit to the US external payments deficit, although it is not clear what the basis is for this claim.

As Mussa said, the truth presumably lies somewhere between these two extreme schools of thought. Mussa suggested that it might lie a little closer to Alfred E. Newman than Chicken Little. Nevertheless, Mr. Mussa concluded:

> Substantial US external deficits probably can and will go on for some time, but the an ever-growing deficit (as a share of US GDP) is not feasible in the long run, and even the present level of the deficit is not likely to be sustained for another five years. The foreign exchange value of the US dollar will need to depreciate substantially, particularly against Asian currencies, as part of the process of adjustment to significantly lower US deficits. This adjustment process will not be completely smooth, but risk of a highly disruptive 'dollar crash' is not particularly great. Policy measures, including more aggressive efforts to improve the US fiscal balance and more rapid appreciation of the Chinese exchange rate, could and should usefully reduce these risks, but the likely benefits of such actions should not be exaggerated.

Mussa said that in order to bring about a substantial reduction in the US external deficit, three things will need to happen over the next decade or so: (1) The US dollar will need to depreciate substantially in real effective terms, probably by at least another 20%,[72] (2) US domestic demand will need to grow more slowly than US output, reversing the trend of the past fifteen years,[73] and (3) domestic demand in the rest of the world will need to grow more rapidly than output. He emphasized that these three basic requirements are not alternatives but rather, they are jointly necessary.

However, US domestic demand for petroleum showed no sign of abatement, prior to the recession of 2008 and we have been engaged in a race to build bigger, more powerful vehicles for some time.[74] The US production of petroleum peaked in the 1970s, and we are increasingly dependent on imports.

One problem with economists is that they almost always seem to be wrong. The economy may be represented as a ship of state; a vessel without a steersman that wanders about in response to winds and currents that are unpredictable. Economists view the ship from afar on land with telescopes, each predicting that it will go this way or that. One might think *a priori* that they should be right 50% of the time just due to chance. But they seem to have an unerring ability to defy the laws of probability and make the wrong prediction. However, Mr. Mussa seems to be more on target than other economists.

Interest rates in the United States also affect the value of the dollar. The Fed has been determined to protect asset bubbles by keeping interest rates low following the stock market crash of 2000–2001. This led to a combined stock-housing bubble from 2002 to 2007. Ben Bernanke, Chairman of the Fed, is sometimes called "Helicopter Ben" because in giving a speech about deflation in 2002, he said that the government can create money and the government can always avoid deflation by simply issuing more money, referring to a statement by Milton Friedman about using a "helicopter drop" of money into the economy to fight deflation. Bernanke also noted that "inflation erodes the real value of the government's debt and, therefore, that it is in the interest of the government to create some inflation."

In 2008, the US Federal Government distributed about $160 billion to taxpayers to stimulate the economy and about a trillion dollars to institutions. Since the US Federal Government does not have this money, it will have to borrow the money, probably mostly from China. The wonder of it all is that China agrees to this—except for the fact that much of that money will be spent on purchasing Chinese goods. The end result will be that China will increase its grip on the US economy. But "Helicopter Ben" will fight deflation.

BANKRUPTCIES

Between 1980 and 2002, 18,000,000 couples and individuals filed for bankruptcy. Typically, more than a million people per year file for bankruptcy. The annual number of bankruptcies is shown in Table 1.20.

The number of personal bankruptcies per year increased dramatically after the early 1980s. The consumer credit industry lobbied Congress for nearly

Table 1.20 Bankruptcy filings in the United States, 1980–2001 (thousands)[75]

Year	Total	Business	Personal
1980	331	44	288
1985	413	71	341
1990	783	65	718
1995	927	52	875
1998	1442	44	1398
2000	1253	35	1218
2001	1106	30	1097
2003	1650	37	1613
2004	1636	36	1600
2005	2078	39	2039
2006	638	20	618
2007	828	28	800
2008	1035 (est)	35 (est)	1000 (est)

ten years in an effort to pass a bankruptcy reform bill. The industry claimed that consumers used bankruptcy as a means of financial scheming, running up huge credit card bills with complete disregard for their ability to repay them, and then discharging them in bankruptcy when they couldn't meet the payments. Finally, on October 17, 2005, Congress passed the Bankruptcy Abuse Prevention and Consumer Protection Act of 2005 (BAPCPA). However, it has been claimed[76] that:

> Abusive filers made up a very small percentage of bankruptcy petitioners. The vast majority of people who file for bankruptcy do so because of huge medical bills not covered by insurance, divorce, job loss, or a death in the family.

As Table 1.20 shows, bankruptcy filings dropped sharply in the aftermath of the BAPCPA. However, the data are skewed by the fact that in expectation that the BAPCPA would be passed, many people in precarious financial condition filed for bankruptcy in advance of passage of the bill. However, the rate of filings was increasing with each month in 2007, and the American Bankruptcy Institute predicts more than a million filings for 2008. Personal bankruptcies for the first quarter of 2008 totaled 230,000.

Democratic US senator from Connecticut Chris Dodd blasted the bankruptcy reform bill, claiming that it benefits banks and credit card companies at the expense of citizens.

WORLD DEBT

As we pointed out in "Rationality of Bankers," there is a long history of defaults on loans by developing nations. According to Delaimaide,

> Latin-American countries began borrowing and defaulting as soon as they gained independence in the 1820s" and "... by 1940 nearly 4/5 of all Latin-American bonds floated in North America were in default.

Yet, that did not inhibit banks from continuing to originate new loans with developing nations. One reason given by Delamaide is that when OPEC tripled oil prices in 1972–1973, the OPEC nations acquired huge amounts of cash, which they deposited in the major banks of the world. The banks didn't have very many customers that could borrow such large sums other than developing nations.

In his 1984 book, Delamaide[77] described the various attempts to deal with wholesale defaults of huge amounts of debt by the developing countries during the financial crisis of 1982–1983. Delamaide opined that these were generally subterfuges of one kind or another:

> There's no solution to the debt crisis, only a resolution. The problem is not debts; but money. The only 'solution' for debts is to get rid of them, either by paying them back or writing them off. The various solutions proposed for ending the crisis presumed that the debtors were capable of honoring their commitments, but that they needed more time to do so than original loan terms allowed. They were countries short of cash, but with the means of generating enough income over time to pay back the debt—was the way this argument ran.
>
> Much of the problem solving by bankers, monetary officials, and governments was so much shadowboxing. They were trying to find a way to keep loan payments current, either by shelling out new money or postponing the date of payment. But their hope was forlorn. The banks cannot prevent losses on their loans, because the losses have already occurred. Most of the shilly-shally in rescheduling loans was simply a shell game to preserve legal fictions. The question was not how to keep the banks from losing money—the money was gone. The problem was to find a genteel way of recognizing that loss in the bank's accounts. The overhang of debt that's on the banks' books can be taken care of in two ways. The banks can write it off and take a corresponding loss. Because of the sums involved, they could not do that right away because they would have been bankrupt. So they played their shell game and postponed the day of reckoning until they had a few years to build up loan loss reserves.

The reality is the same: The losses have already occurred, whether this is acknowledged now or later.

The other way to take care of these losses is to inflate the debt away. This wipes out the loss by wiping out the value of money. It is effective in purifying the banks' books, but obviously has other costs, It not only purifies away the banks' bad loans, but the well-intentioned savings of the rest of us.

The US administration and most American bankers rejected the need for grandiose solutions. They maintained that the problem was a temporary liquidity crisis and did not require a fundamental solution. The crisis originally arose from an extended recession. The obvious way to overcome the difficulty, then, would be an extended period of growth. Simple. Perhaps the only hitch was the experience of the previous decade which demonstrated decisively that all the king's horses and all the king's men seemed powerless to move the economy one way or the other, regardless of what policy was decided upon. So this school of thought boiled down to keeping your fingers crossed that growth would resume and continue, and then the debt problem would disappear.

DEPOSIT INSURANCE

According to the Federal Deposit Insurance Corporation (FDIC) official website:

> The Federal Deposit Insurance Corporation (FDIC) preserves and promotes public confidence in the US financial system by insuring deposits in banks and thrift institutions for at least $100,000; by identifying, monitoring and addressing risks to the deposit insurance funds; and by limiting the effect on the economy and the financial system when a bank or thrift institution fails. . ...
> Since the start of FDIC insurance on January 1, 1934, no depositor has lost a single cent of insured funds as a result of a failure. The FDIC receives no Congressional appropriations—it is funded by premiums that banks and thrift institutions pay for deposit insurance coverage and from earnings on investments in US Treasury securities. With an insurance fund totaling more than $49 billion, the FDIC insures more than $3 trillion of deposits in US banks and thrifts—deposits in virtually every bank and thrift in the country.[78]

The above-cited paragraph from the FDIC official webpage indicates that the FDIC holds an insurance fund of $49 billion to cover more than $3,000 billion in deposits, for a 1.5% ratio. This appears to be appropriate for "normal" times but it certainly would not be adequate for a financial calamity of the magnitude of the S&L scandal.

In order to receive this benefit, member banks must follow certain liquidity and reserve requirements. Banks are classified according to their "risk-based capital ratio." When a bank becomes undercapitalized the FDIC issues a warning to the bank. When this ratio drops below 6% the FDIC can force a change management and require the bank to take other corrective action. When the bank becomes critically undercapitalized the FDIC declares the bank insolvent.

The history of the FDIC is aptly described in a FDIC document.[79]

> Thousands of banks failed in 1933 and never reopened. The confidence of the people still was shaken, and public opinion remained squarely behind the adoption of a federal plan to protect bank depositors. Opposition to such a plan had been voiced earlier by President Roosevelt, the Secretary of the Treasury and the Chairman of the Senate Banking Committee. They believed a system of deposit insurance would be unduly expensive and would unfairly subsidize poorly managed banks. Nonetheless, public opinion held sway with the Congress, and the Federal Deposit Insurance Corporation was created three months later when the President signed into law the Banking Act of 1933. The final frenetic months of 1933 were spent organizing and staffing the FDIC and examining the nearly 8,000 state-chartered banks that were not members of the Federal Reserve System. Federal deposit insurance became effective on January 1, 1934, providing depositors with $2,500 in coverage, and by any measure it was an immediate success in restoring public confidence and stability to the banking system. Only nine banks failed in 1934, compared to more than 9,000 in the preceding four years.

This document goes on to say:

> In its seventh decade, federal deposit insurance remains an integral part of the nation's financial system, although some have argued at different points in time that there have been too few bank failures because of deposit insurance, that it undermines market discipline, that the current coverage limit of $100,000 is too high, and that it amounts to a federal subsidy for banking companies. Each of these concerns may be valid to some extent, yet the public appears to remain convinced that a deposit insurance program is worth the cost, which ultimately is borne by them. The severity of the 1930s banking crisis has not been repeated, but bank deposit insurance was harshly tested in the late 1980s and early 1990s. The system emerged battered but sound and, with some legislative tweaking, better suited to the more volatile, higher-risk financial environment that has evolved in the last quarter of the 20th century.

During the late 1940s and 1950s there were no more than five bank failures in any single year. Fewer than 10 banks failed per year in the 1960s. Because most of the banks that failed during the period 1942–1970 were small institutions, insurance losses remained low. In just four of these years did losses exceed $1 million, and losses averaged only $366,000 per year. However, the low incidence of failures was regarded by some as a sign that the bank regulators were overly strict, operating with policies and practices rooted in the banking crises and economic chaos of the 1930s. In 1963, Wright Patman (Democratic Chairman of the House Banking and Currency Committee) said there should be more bank failures and we have gone too far in the direction of bank safety. Looking at the bankers of the 1980s and the 2000s, it appears that Mr. Patman got his wish fulfilled—and then some!

The new generation of bankers who came to power in the 1960s abandoned the traditional conservatism that had characterized the industry for many years. Instead, they began to strive for more rapid growth in assets, deposits and income by taking greater risks. They were aided and abetted by liberalization of regulations at the state and national levels. The size of bank failures increased in the 1970s but the losses were not beyond the capability of the FDIC. However, the housing bubble and savings and loan scandal of the 1980s brought on much more extensive losses. From 1982 through 1991, more than 1,400 FDIC-insured banks failed, and 131 remained open only through FDIC financial assistance.

The Federal Savings and Loan Insurance Corporation (FSLIC) is a now-defunct institution that once administered deposit insurance for savings and loan institutions in the United States. It was abolished in 1989 by the Financial Institutions Reform, Recovery and Enforcement Act, which passed responsibility for savings and loan deposit insurance to the Federal Deposit Insurance Corporation (FDIC). The savings and loan scandal of the 1980s is discussed in "The Savings and Loan Scandal of the 1980s" in Chap. 2. More than 1,000 savings and loan institutions (S&Ls) failed in "the largest and costliest venture in public misfeasance, malfeasance and larceny of all time."[80] The FSLIC insurance fund which amounted to a few $ billion, was grossly inadequate to deal with the $160 billion cost of bailing out the S&Ls.

One of the most notable features on the landscape of the banking crises of the 1980s was the crisis involving Continental Illinois National Bank and Trust Company (CINB) in May 1984, which was the largest bank resolution in US history prior to 2008. As the nation's seventh largest bank, Continental forced regulators to recognize not only that very large institutions could fail, but also that bank regulators needed to find satisfactory ways to cope with such failures.

The sub-prime mortgage fiasco resulted in eight bank failures during the first half of 2008. Notable among these was the failure of IndyMac that will cost the FDIC something like $8 billion. According to a CNN News Release, "76 banks are currently under scrutiny" and "regulators are bracing for 100–200 bank failures over the next 12–24 months." By January 2009, every major bank in the US required a federal bailout.

After enduring the sub-prime mortgage crisis of the early 21st century, several things are clear:

- Banks can act (and have acted) with abandon to invest funds in speculative ventures, knowing that depositors will not lose their money because the FDIC will bail them out if they fail.

- Contrary to the opinion of Mr. Patman, the S&L scandal of the 1980s and the sub-prime scandal of the 2000s underscore the need for banks to be regulated more closely and deprived of speculative options.

- Deregulation is not the same as no-regulation.

- The FDIC has adequate funds to cope with the year-to-year occasional bank failures that occur in ordinary business.

- The FDIC is not equipped with funds or manpower to deal with widespread banking abuses such as occurred in the S&L scandal of the 1980s or the sub-prime fiasco of the 2000s.

- The very existence of a FDIC backed by the government should be predicated on the requirement that regulators must prevent wild, speculative ventures by banks using depositors' money.

For the future, if short-term interest rates rise inordinately, banks and investors holding long-term fixed mortgages will be back in the same situation that S&Ls were in during the 1980s. Revenue from fixed-rate mortgages will not compensate for interest paid to depositors. Perhaps that is why the Fed continues to drive down interest rates. If interest rates rise too much, the whole banking business could collapse.

REGULATION, DEREGULATION AND NO REGULATION

It is widely believed that utilities played a significant role in the stock market crash of 1929. As the nation sunk into a deep depression in the early 1930s with thousands of bank failures, it became apparent that there was a need for

Governmental regulation of banks and utilities. Legislation was passed in the 1930s to provide regulation of banks and utilities, and later, the transportation industries were regulated as well. Government regulation of banks was supposed to provide oversight to enforce conservative investment practices, with depositors' accounts insured by the Government. Government regulation of utilities required that in consideration of the exclusive monopoly provided to a utility in its locale, the utility must operate for the benefit of the public it served, although it was entitled to a fair profit—set by Government appointed regulatory agencies. The regulation of the transportation industries was aimed at assuring public safety and fair pricing.

In the late 1970s, it became in vogue for economists (particularly those leaning toward the Republican view) to argue that Government regulation was a stifling influence that inhibited progress. According to this view, introduction of competition in these industries would foster innovation and progress, leading to improved service and lower rates to the public. In addition, there was a growing sympathy for the Republican view that nothing the Government does is good, and minimization of all Government activities (except for the military) became a central theme of the Republican Party.

By 1980, there was a widespread belief that deregulation of formerly regulated industries would provide great benefits. In 1980, Ronald Reagan was elected president. He had an unwavering antithesis to any Government activity at all (except of course the military) and with a religious fervor, pursued deregulation of everything. Furthermore, he interpreted deregulation as *no regulation.*

It is remarkable that during the presidential primary of 2008, all three major Republican presidential candidates vied with one another in the claim that they were the most like former president Ronald Reagan. Their awe and reverence for Mr. Reagan was limitless.

Two very good things happened during Reagan's term in office, *neither of which was due to his actions or policies.* One was the sharp drop in oil prices and the temporary end of tight energy supplies, and the other was the collapse of the Soviet Union and the end of the cold war.

However, two very bad things occurred during the Reagan administration, and these *were* due to his overt policies.

One was that he originated the new fiscal policy that has guided the Republican Party from 1980 through 2008. While the Republican Party in the post-WWII years favored reduced government expenditures, lower taxes, a balanced budget, reduced foreign aid, and a foreign policy that leaned toward isolationism, Mr. Reagan made it all seem so simple: Cut taxes, particularly for the wealthy, and the trickle-down effect would bring prosperity to all Americans, and governmental revenues would rise (not fall) due to the putative resultant prosperity. At the same time, Mr. Reagan was unable to cut government expenditures, and in fact he increased defense expenditures, so that he originated the era of annual multi-hundred billion dollar deficits after 1980. His claim that lower taxes produce greater government revenues, which has been a Republican mantra for 28 years, has been proven wrong ever since. The one exception was the peak period of the dot.com bubble that generated so much temporary capital gains tax that deficits were greatly reduced—until the bubble popped.

The other innovation introduced by Mr. Reagan was the belief that banks should not only be deregulated but they should be completely unregulated. The S&L scandal of the 1980s occurred under his administration, with his approval, both tacit and overt, as administered by his Secretary of the Treasury, Donald Regan.

According to Delamaide:[81] in Reagan's third fiscal year, "the deficit had turned into a mad bull elephant, raging out of control."

Not only did the projection [of the deficits] give a new dimension to deficit spending, but doubling the original projections marked a new tack in the numbers game played by Reagan's budgeteers.

During the presidency of Ronald Reagan, multiple scandals developed which resulted in a number of administration staffers being convicted. The most well known, the Iran-Contra affair, involved a plan whereby weapons were sold to Iran and the profits diverted to fund the Nicaraguan Contras, in violation of US and international law. This was done because Congress would not authorize funding the Contras from government funds. For this alone Mr. Reagan should clearly have been impeached.

A total of 225 people who served in the Reagan administration either quit, were fired, arrested, indicted, or convicted for either breaking the law or violating the Ethics Code; Edwin Meese alone, the Attorney General, was investigated by three separate Special Prosecutors.

Several other controversies also occurred in the Reagan administration; one involved Department of Housing Secretary Samuel Pierce and his associates. Wealthy contributors to the administration's campaign were rewarded with funding for low income housing development without the customary background checks, and lobbyists, such as former Secretary of the Interior James G. Watt, were rewarded with huge lobbying fees for assisting campaign contributors with receiving government loans and guarantees. Six administration staffers were convicted.

Also involving the EPA: funds from the Superfund to clean up toxic waste sites were released to enhance the election prospects of local politicians aligned with the administration.[82]

As we discuss at length in " How Mr. Reagan Made a Bad Problem Worse" in Chap. 2, Reagan's policies of no regulation was a major factor contributing to the S&L debacle of the 1980s. Nevertheless, deregulation as a concept continued to flourish, even in the face of its continual abject failures. Deregulation of telephone service has produced higher rates and poorer service. Prior to deregulation of telephone service, I had one telephone book published by AT&T, and it contained everything I needed. Today, I have eight telephone books and none of them are worth opening. Deregulation of airlines has sent many airline companies into bankruptcy and airline service gets worse and worse. Deregulation of banks produced the sub-prime mortgage fiasco of 2002–2007. With the Government under-funding regulatory agencies, even the low-level regulation appropriate for a "deregulated" system has been absent. As JKG said:

> In recent times it has become obligatory for the [bank] regulators at every opportunity to confess their inadequacy, which in any case is all too evident.[83]

PENSION PLANS

CORPORATE PENSIONS

DEFINED BENEFIT PLANS

Traditionally, large corporations have provided pension plans for their long-term employees at no cost to employees. These *defined benefit plans* typically required up to 10 years for vesting, and promised to pay out some percentage of an employee's maximum earnings based on their years of service. Most plans began retirement payments at age 65.

Shiller pointed out that most *defined benefit plans* were not indexed to inflation and that people who retired and lived a long time after retirement under defined benefit plans often saw a substantial part of the real value of their pensions eroded away by inflation. However, most of these plans were based on the highest salary achieved during the worker's tenure, so they were implicitly indexed for inflation for the period of employment. But, they were not indexed for inflation after retirement.

94

It is not clear what requirements (if any) were imposed in the past (prior to ERISA in 1974) on corporations to maintain funds for future retirement payments. It seems likely that some companies voluntarily put aside funds to pay for future retirement obligations, while others may have merely hoped that future earnings would expand to cover future retirement obligations. On the other hand, some companies treat future retirement payments as a probability, rather than a certainty. For example, this quotation is taken from the Kaiser Permanente rulebook for retirement of doctors:

> Benefits from this plan are based upon Health Plan's ability to pay. No trust or other separate fund or individual account is established, nor is an annuity established for the plan participants. If Health Plan is unable to pay its obligations, Plan participants are considered general creditors and have no preferred status or priority over claims of other Health Plan creditors.

Even for those companies that did put aside funds for future retirement obligations, there are questions as to how much was needed on an actuarial basis, and what assumptions should be made regarding future earnings of the funds. While investment in government bonds may be the most prudent course, such stodgy investment policies fell out of favor beginning with Reagan's election in 1980. Under ERISA of 1974, minimum funding requirements were established for defined benefit plans, but retirement plans were permitted to assume corporate bond yields for future earnings (rather than lower government bond yields). With the advent of the great stock bull market in 1982, corporations tended to invest retirement funds into the stock market. In some cases (Enron, Ford, ...) they invested the retirement funds almost exclusively into their own corporate stock (thus providing a buying boost for their stock) but when their stock price went south, the retirement assets also declined.

Legislation has been enacted to protect the interests of employees with defined benefit plans. These include the Employee Retirement Income Security Act (ERISA) of 1974, the establishment of the Pension Benefit Guaranty Corporation (PBGC) of 1974, and the Pension Protection Act of 2006. Despite these small steps toward pension security for employees, significant problems remain for those companies that either (a) made inadequate provisions in the past, (b) made poor investment decisions in the past, or (c) face financial hardship in providing for promised retirement benefits. The PBGC does not seem to have adequate funding to protect the interests of employees with failing corporate *defined benefit plans* and as a result, it has limited its responsibility to a maximum of about

$4,000/month at age 65, or about $2,500/month at age 58 and does not provide health benefits.

The PBGC used to publish an annual list of the 50 companies with the most under funded pension plans. But in 1997, under pressure from companies, they ended this practice. There are conflicting data on the Internet. According to an Internet source, status reports for 2004 for 1,108 pension plans covering about 15 million workers and retirees were filed with the PBGC by April 15, 2005, showing that under funded pension plans reported a record shortfall of $353.7 billion, up from $279.0 billion for 2003. The under funded plans had $786.8 billion in assets to cover more than $1.14 trillion in liabilities, for an average funded ratio of 69%. It was claimed that as of September 30, 2004, the PBGC estimated that the total shortfall in all insured pension plans exceeded $450 billion. In the same report it said that a loophole allowed United Airlines to go for years without making any cash contributions to its retirement plans, without paying additional premiums to the PBGC, and without sending under-funding notices to plan participants even though United's plans have an aggregate funding shortfall of almost $10 billion and an average funded ratio of 41 percent.

Recent data show that the underfunding of corporate pension plans was a maximum in 2002 and had greatly diminished by 2007. Since a significant portion of these funds is invested in stocks, the deficit or surplus (as the case may be) will depend on stock market performance. The surplus built up in the dot.com boom in 1997–1999 disappeared in the aftermath of the stock market collapse. The resurgence since 2002 has been due to an increasing stock market. It seems likely that with the decline in stocks in 2008, that the surplus reported for 2007 in Table 1.21 might disappear.

DEFINED CONTRIBUTION PLANS

An alternative to a *defined benefit plan* is a *defined contribution plan*. In a *defined contribution plan* the employee makes a contribution of funds from each paycheck to a vested account in his name and in most cases, the employer also makes a contribution. These funds are maintained by the employee in a separate account, administered by a large financial organization such as Fidelity or Principal.com, and are beyond the reach of the employer. Most of these accounts are so-called 401(k) accounts, but there are also 403(b), Keough and other forms of *defined contribution plans*. Some characteristics of these plans are compared to defined benefit plans in Table 1.22. In general, there are great advantages to *defined contribution plans*. These include (1) the funds are under your management and cannot be lost by the company, (2) you can up the ante by raising your contributions voluntarily, (3) in an emergency, you can withdraw funds any time but pay a

Table 1.21 Funding for single-employer pension plans for S&P 500 companies (in billions of dollars)[84]

Year	Pension assets	Pension liabilities	Pension surplus
2007	1,505	1,441	63
2006	1,471	1,511	−40
2005	1,318	1,458	−140
2004	1,265	1,430	−164
2003	1,113	1,278	−165
2002	951	1,169	−219
2001	1,090	1,087	3
2000	1,239	1,013	226
1999	1,274	994	280
1998	1,144	1,018	126
1997	991	794	197

Table 1.22 Comparison of defined benefit plans with defined contribution plans

Characteristic	Defined benefit plan	Defined contribution plan
Who pays for it?	Employer	Employee and (usually) employer
Minimum age for retirement	Set by employer, typically 65	Set by employee, but 10% penalty if withdrawals made prior to age 59.5
How secure is it?	Dependent on funds set aside by employer and health of company.	Absolutely safe, but amount is dependent on financial management by employee
Does it provide for heirs if you die early?	Typically, no benefit once you die.	Yes. The funds are always there for your heirs, even if you die.
Can you lose it?	Yes, if a company mismanages the fund, or goes into financial hard times.	Only if the employee mismanages the funds.
Can you increase the amount by adding more contributions?	No. The plan is fixed.	Yes, up to the statutory limit. But it reduces your paycheck.
Is it portable?	If you change jobs, the benefits may not be vested, and if they are vested, the value may be small.	If you change jobs, you retain ownership of the 401(k) account.

10% penalty if before age 59.5, and perhaps most important of all, (4) your heirs inherit the fund if you die before you use it up—it is your money, not the employer's.

During the latter part of the 20th century, there was a gradual shift away from *defined benefit plans* to *defined contribution plans*, and today, the majority of retirement plans are 401(k) *defined contribution plans*.

The main disadvantage to the employee of a *defined contribution plan* is that it requires that the employee be sufficiently astute financially to invest his or her funds wisely over many years. The common wisdom is that stocks represent the best long-term investment, and indeed that has been true since the great bull market started in 1982. Most white collar and professional employees put the majority of their 401(k) funds into stocks and live or die with these investments. Shiller pointed out:

> ... 401(k) and similar plans were designed to give ordinary people economic security in retirement by encouraging them to mimic the portfolio strategies long pursued by the wealthy. ... But little attention is usually paid to the fact that the wealthy, because of the overall level of their assets, have less reason to worry about losing substantial amounts in a market decline.

Thus 401(k) plans free employees from tyranny by employers, but they place the additional burden of responsibility on the employee for investment decisions. Again, as Shiller pointed out, employees are typically given several investment choices, including many varieties of stock market investments, and "thus there is a not-so-subtle nudge in the direction of investing heavily in the stock market." Shiller went on to say that those who offer 401(k) plans typically provide many options for investing in stocks but very few (if any) bond options.

From the employers' point of view, the defined contribution 401(k) plan was very attractive. It eliminated long-term liability by the company. As Shiller said:

> Employers promised only that they would contribute a certain amount to an employee's nest egg while he was working. What happened to the money after they parted was the employees' responsibility. How long they lived, and how far their savings stretched, was their problem, not the employer's.

As Mahar pointed out,[85] boomers could be expected to live longer than their parents, and corporate profits were sluggish. In addition, the Employee

Retirement Income Security Act (ERISA) passed in 1974 made it both more difficult and more expensive to run a traditional pension program.

Shiller estimated that 2/3 of 401(k) funds were invested in the stock market in 2003, and with the rise in market average from 2003 to 2006, that proportion probably increased significantly. The downturn in late 2007 and early 2008 made retirement schedules precarious for many middle-aged employees. Shiller emphasized this risk. He also pointed out: "Managers of 401(k) plans generally do not offer advice to employees about how they should make their allocations." Worse still, managers are not held responsible for the choice of investments offered to participants.

In the 1980s, the California Institute of Technology (Caltech) offered participants in its 401(k) plan the choice of TIAA-CREF or Mutual Benefit Life Insurance (MBLI). TIAA-CREF offered bond and stock investments and MBLI offered only a bond-like investment. MBLI was a triple-A company for many years, but then fell prey to "go-go" real estate ventures of the 1980s and its finances declined sharply toward the end of the 1980s. Caltech's view (apparently) was that having chosen MBLI initially, it was under no further obligation to track the company and its status and performance. Yet, there is some evidence that the decline of MBLI was known to many on Wall Street, although Caltech was more concerned with the theory of relativity. When Caltech finally woke up to the danger, it was rather late in the game, and most of the 401(k) participants in MBLI were stuck there, as MBLI receded into bankruptcy. Caltech refused any relief to its employees, and their position was upheld by the courts. In the final settlement, employees were given a multi-year payout settlement that involved significant losses to all participants compared to what they would have accumulated had they been able to transfer their funds to TIAA-CREF. Considering that Caltech chose MBLI in the first place from hundreds of companies, should Caltech have had the responsibility to track MBLI, and warn employees to transfer out at a sufficiently early date? Apparently the courts said no.[86]

THE PUBLIC SECTOR

According to Shiller:

> State and municipal pension plans face an unfunded liability of upwards of $1 trillion. And the worse news is the public sector has an additional unfunded liability half again as big for other post-retirement expenses such as health care, a staggering $1.5 trillion.

According to the New York Times:[87]

> Almost half of the states have been under-funding their retirement plans for public workers and may have to choose in the years ahead between their pension obligations and other public programs. . ... All together, the 50 states have promised to pay some $2.7 trillion in pension and retiree health benefits over the next 30 years. This amount does not include separate retirement plans run by local governments. While some states are managing their costs reasonably well, the center found that others, like New Jersey and West Virginia, have made serious mistakes and are now cutting education and health programs as they struggle with costs incurred decades ago. Still more states are at risk of being caught in a similar squeeze, . . . because they are not setting aside enough money now, as their populations age and more public employees approach retirement. . ... Unlike companies, state and local governments are not subject to federal pension laws, which set uniform standards for private industry. If a company skips its required pension contributions, it can be required to pay a big excise tax. No comparable enforcement mechanism exists for states.

There are some extreme cases that have made the news.

According to Internet sources, New Jersey decided in 1994 to stop setting aside money in a fund to pay for health care for its retired public workers, thus allowing a big tax cut. Public workers were told that as long as they worked 25 years, the system would provide virtually free health care for them when they retired, often when they were as young as 55. It is claimed that New Jersey will need about $58 billion, in today's dollars ($6700 per capita), to provide all the care it has promised its current and future retirees. That's nearly twice the state budget and nearly twice the amount of its outstanding debt. However, the Governor claimed[88] that the debt is more like $32 billion, or $3,700 per capita. But he did admit that: "those numbers will grow dramatically in the years ahead if we accept the status quo." In addition to the bonded debt, he added $25 billion in unfunded pension liabilities and an estimated $60 billion in future health care costs for retirees. He concluded that the total obligation of the State of New Jersey amounts to $45,000 per household.

And, because of the step it took in 1994, the state has virtually no money in reserve to cover those costs.

In addition, New Jersey's towns and other local governments owe about $10 billion for health care for their own retirees.

In a similar manner, the Orange County Register reported that:

> In 1996, the city of San Diego purposely began under-funding its municipal pension system even as it increased retiree benefits, a policy that continues even today. For years, above-average returns on Wall Street permitted the city to convince the trustees of the retirement system that this approach was less dangerous than it sounded. When the financial markets went down beginning in 2000, city officials kept the trustees on board by increasing pension benefits yet again and by creating special benefit enhancements that seemed targeted toward the trustees and the leaders of key municipal unions. In other words, the city promised more, put aside less and hoped that the financial markets would come to the rescue once again. The markets didn't cooperate, however, so now the city retirement system has a $1.4 billion deficit and hundreds of millions of dollars more in unfunded retiree health-care costs.[89]

It has been reported on the Internet that the total Illinois bond and pension debt amounts to $65 billion, or about $5,300 per capita.

These are just a few examples of many states and municipalities that face serious obligations in the future that they are unlikely to be able to meet without great privation.

THE VALUATION OF COMMON STOCKS

Valuation of common stocks was discussed at length by Smith.[90] In the early 1920s, stocks were valued primarily by the dividends that they paid. Since stocks were riskier than bonds, stocks had to pay a higher dividend than bonds to compensate for the higher risk. Typical dividends on NYSE listed industrial stocks were in the range 5–6%. In those days it was expected that an industrial corporation would pay out roughly half of its earnings as dividends. Hence if dividends were 5% of the stock price, earnings would be 10% of the stock price, so the "normal" stock price/earnings ratio (P/E) would be about ten, and that was a common *rule of thumb* for valuing common stocks.

In those days, the main reason to own a stock was to share in corporate profits by collecting dividends, and the current dividend, together with prospects for further increases, dictated the current price of a stock.

However, some companies began to pay out a lower percentage of earnings for dividends, using a higher proportion of retained earnings for expansion. A number of prominent voices in the early 1920s argued for new approaches to valuing stocks based on expectations of future growth in earnings that would allow implicit future increases in dividends. Thus, in an era of growth the argument went that the price of a stock should reflect future earnings prospects more than current dividends. As we have discussed previously, the advent of the automobile, travel, transportation by truck, industrialization, and consumer spending as major elements of the economy, spurred investors' imaginations for future growth. In an era where it was perceived that companies might grow enormously in a short time, a belief grew that a higher P/E ratio was justified. As can be seen from Figure 1.10, the average P/E ratio of the S&P 500 Index rose upward from below 10 to just over 30 in the roaring twenties, just before the crash of 1929.

In the current era of the early 21st century, the philosophy of valuation of stocks has been inverted. Instead of valuing stocks in terms of their ability to pay dividends, and treating stocks as being riskier than bonds, therefore requiring a higher yield than bonds, the markets have adopted a very different paradigm. Today, we have a "go-go" bubble mentality, and the greater risk is seen in bonds—the risk that you will miss out on a meteoric rise in stock prices. There are two possible reasons to buy stocks—for their dividends, or for price appreciation. In the current markets, dividends provide little attraction, and price appreciation seems to be the overwhelming motivation for owning stocks. Driven by the great

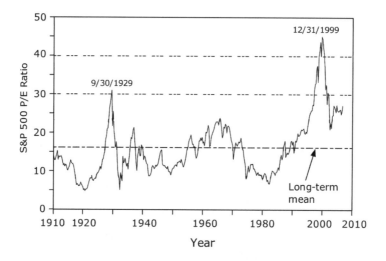

Figure 1.10 History of the price/earnings ratio of the S&P 500 Index. (Adapted from: "A History of the Global Stock Market from Ancient Rome to Silicon Valley," by B. Mark Smith).

rise in stock prices since 1982, many investors expect and count on double-digit increases in stock prices each year and sneer at merely collecting interest from bonds or dividends from staid old blue-chip stocks. In fact, dividends are hardly a consideration in determining stock prices in 2008, as evidenced by the fact that the dividend yield of the S&P 500 in 2007 was only 1.8%, and 0.5% on NASDAQ 100 stocks. However, the only way that one can directly share in the profits of a corporation is through dividends. It is only because of the expectation that one can sell shares to a future buyer that stock holders are willing to bid up stock prices and thereby drive down dividend yields. Some companies, such as Apple Computer—that make outstanding products—must plow back all of their earnings into new product development. They have never paid a dividend and may not pay one for a very long time, if ever. As Eliot Janeway said: "The price of staying in business is continuous investment".

There are many theories as to how to value common stocks. Most analysts concentrate on individual stocks, and talk about earnings, earnings prospects, takeover prospects, sales, etc. The ability of analysts to predict future stock movements seems to be minimal (see "The Valuation of Common Stocks" on "Innocent Fraud"). On the other hand, those with inside information have a special advantage, and even though the SEC purports to constrain use of inside information in stock trading, there are thousands of instances of major stock movements prior to public announcements of great importance. This suggests that insider trading, though illegal, is rampant. Another theory is that the whole stock market tends to move together, upward and downward, and picking individual stocks is an effort in futility, except for a few special situations. Hence, market timing is the key to stock gains according to this theory. Indeed, if you buy stocks at the wrong time, it can take many years to break even. According to some, the stock market is implicitly a money market. When the money supply is plentiful, money flows into the markets and stocks rise. When the money supply is tight, money flows out of the markets and stocks fall. That is why investors respond with exuberance to each action of the Federal Reserve System. The Republican administrations since Reagan in 1980 passed tax legislation favorable to stock ownership and have flooded the money markets, making savings less attractive. From this viewpoint, the P/E ratio is not very important. However, as Figure 1.10 demonstrates, the long-term mean of the S&P 500 P/E is 15.7, and when this ratio reaches upward toward 30 or 40, the markets inevitably seem to crash afterwards.

But once the public discarded dividends as the basis for valuing stocks, the valuation process changed from being objective to being subjective. The principal reason to own stocks is no longer to share in the profits from companies via dividends, which tend to change gradually with time. Impatient investors want much larger returns, more quickly. The main reason that people have for investing in stocks is to hold paper that they hope to sell at an inflated price in the future. How high (or low) a stock price can go is no longer a matter of computation based on objective standards, but instead is a matter of whim, fancy, hope, expectation and luck. The result has been an increase in volatility. However, fortunately for investors, there is a widespread belief throughout the land that stocks are the best intermediate-term and long-term investments, and the great majority of 401(k) retirement funds are invested in stocks. The continued expansion of 401(k) retirement plans has fed a seemingly endless supply of new money into the stock markets since about 1982, that has fueled and propagated this subjectively valued market onward and upward[91] —except for a few notable collapses (1987, 2000–2001, and 2007–2008). This change from objective to subjective valuation of stocks is similar in some ways to the transition from representational art to abstract art that occurred in the 20th century. The world of art discarded the traditional values and standards, and curators, museum officials, gallery owners and critics now routinely endorse art that defies logic, sensibility, esthetics and plain common sense, while phonies, nonentities, and flim-flam artists are routinely pronounced to be great geniuses of our time. This is not unlike the situation with common stocks, for we live in an era of deregulation and no regulation, and we have no standards for the valuing stocks or art tied to any tangible criteria, but rather, we depend on a herd-mentality of hope and expectation, urged on by so-called "experts." The world of classical music has gone through a similar transition.[92]

It is perhaps noteworthy that Smith[93] described this process of the transition from objective to subjective valuation of stocks as if it made sense and represented progress. For example, Smith quoted Greenspan as describing the dot.com boom as follows:

> There is at root here something far more fundamental – the stock market seeking out profitable ventures and directing capital to hopeful projects before the profits materialize. That's good for our system. And, in fact, with all of its hype and craziness, is something that, at the end of the day, probably is more plus than minus.

Greenspan also indicated a factor that is important here, namely that people are willing to pay a large premium for a small chance to win a really big payoff. He called this the "lottery principle." Thus, if you have a lottery with a one-in-a-million chance of winning a million dollars, the nominal value of a lottery ticket would be one dollar. But lotteries typically give you a one-in-ten-million chance to win a million dollars, and yet the public is not dissuaded from buying the tickets. This effect plays into the current psychology of the stock market. Many investors are not interested in the slow growth of big-cap stocks and would rather invest in a selection of emerging stocks, on the belief that if just one of them "becomes the next Microsoft" it will overcome losses from the others.

Smith claimed to have:

> ... demonstrated how standards of valuation for American stocks evolved progressively over the twentieth century, from the more conservative to the more liberal. Beginning in the 1920s, investors started to look to future earnings growth rather than simply current dividends as the source of value in equity investments. By the late 1950s, ... dynamic valuation methodologies based on earnings growth had **triumphed** over static approaches relying solely on current dividends. (emphasis added)

This paragraph seems to imply that valuation of stocks had progressed and *triumphed* by placing value on putative future earnings. However, that is not the way it works in reality. Most investors don't have a clue as to future earnings, and even experts have shown a notable inability to predict them. Stocks are valued not on "future earnings growth" which is far beyond the ability of economists to predict, but rather on future growth in P/E ratio, which depends not on corporate performance but human herd mentality in the markets. Maggie Mahar[94] described how investment veterans such as Morgan Stanley's Byron Wein "seemed out of touch" where "the whole concept of fundamental or intrinsic value ... had become a pejorative term." She described 64-year-old Wein responding to a young analyst who was advocating a stock priced at more than 100 times earnings: "How do you arrive at your valuation? Show me the parameters you're using." According to Mahar, "The young analyst just stared at the 64-year-old market strategist." Mahar emphasized: "... that by 1999, a corporation's assets, its cash flow, and even its revenues had little relevance to the total value investors were willing to assign to it."

Smith also claimed that as the 20th century progressed, "investors became more and more willing to take on the risks, and stock prices rose accordingly." On the contrary, there is no evidence that investors were aware of risks and were willing to take them. More likely, there was a widespread belief that the markets could only go up, and if there were a short-term hiccup, it would speedily recover. Although, in reality, investors have taken big risks (like buying Bear-Stearns at $170/share in January, 2007)[95] they did not have an inkling that they were taking a risk, and thought they were secure. Indeed, the mentality of the public and financial reporters in the late 20th century and beyond is that it is normal for markets to advance at double-digit rates while it is grossly abnormal and incredible for markets to go down. So we have this strange dichotomy. The public is unwilling to take major risks, but does so unwittingly in the belief that their investments are secure in a market that only goes up. That is why it was so cataclysmic when the real estate market went south in 2007–2008; all those people who borrowed trillions were convinced that it couldn't go down. And when it does go down, the public can hardly believe it, and demands that the government must bail them out, so the Federal Reserve obeys that demand as "Helicopter Ben" drops money on the banking system.[96]

A question on the minds of many people is whether stocks are a good long-term investment. In discussing this, it is common to trace out the histories of major market averages like the S&P 500 and the DJIA. However, these indices are constantly weeding out weak members and replacing them with stronger prospects, thus providing a rosier picture of long-term stock prices than a constant portfolio. Over the past, bull markets in the stock market have outrun earnings gains driving up price/earnings ratios (P/E) to high levels until they were unsustainable, and inevitably, crashes resulted. Such peaks in P/E were reached in 1901, 1929, 1964 and 2000 (see Figure 1.10).

After the 1901 peak there was a bear market from 1901 to 1920 in which the real S&P index stayed flat with little gain. The bull market of the 1920s drove the real S&P from 70 to 290, a gain of 400%. The ensuing bear market drove the real S&P down to the 120-range (including a number of dips well below 100) where it languished for about 20 years until a new bull market was born around 1950. The real overall S&P index was flat from 1900 to 1950 (except for a strong peak in the 1920s and a precipitous drop in the early 1930s) that was sandwiched in this flat period. Figure 2.3 in "The Bull Market of 1982–1995" in Chap. 2 shows data on the Standard and Poor's composite stock index. The "real" index is adjusted for inflation. Be aware that the lower graph is logarithmic so vertical variations are compressed. The bear market from 1967 to 1981 dropped the real S&P by 60% and it took until 1987 to recover to the level of 1967, a 20-year period. The greatest bull market of all ran from 1982 to 1999 during which the real S&P reached about 1300. At its height, the P/E was over 40. A major correction downward followed in 2000–2001 in which the real S&P dropped to about 900 and the P/E dropped to about 24–25. But even the drop in the P/E from 40+ to 25 left the P/E in very high territory and it remained susceptible to a major correction. Despite this, there was a recovery and the S&P and the P/E rebounded. Whenever the S&P P/E ratio goes into the stratosphere, an extended bear market is likely to follow. As a long-term investment vehicle, investing in stocks is great if your timing is right, and if your timing is wrong, it can take 10, or 20 or more years to recover. In early 2008, the preponderance of evidence was that the markets were precariously high. This, of course, is based on historical precedent. There is no proof that history always repeats itself. But the risk was great.

There have been nine major bear markets since 1950, in 1957, 1961, 1966, 1968, 1973, 1980, 1987, 1990 and 2000. Declines in these bear markets ranged from 21% to 49%. The fact that stock prices can vary so widely over relatively short durations shows that valuation of stocks is inherently a subjective process, controlled more by herd behavior than by rational economic analysis.

There is a theory regarding a putative inverse relation between P/E and interest rates. It goes something like this... If the P/E were say 20:1, it would imply that the company is making 5% on your investment and that ought to compete with a 5% interest rate. Similarly when, interest rates are 4% that would allow P/E to be 25:1. However, the interest is on a government bond that is relatively safe compared to a stock. Furthermore, only a fraction of the "E" is returned to you in the form of dividends. When long-term interest rates and P/E ratios are plotted on the same axes, there is some (but certainly not consistent) correlation between interest rates and P/E. In fact, the highest interest rates that ever occurred were in the 1980s when the P/E was near an all-time low. Obviously, fixed interest investments can compete with stocks when the interest rate rises. But there have been many long periods for which both interest rates and P/E were low, or were both high.

INTERNAL FEEDBACK AND ENDOGENOUS RISK

A paper on the Internet[97] provides a very good analysis of internal feedback mechanisms that can exaggerate price movements for assets, resulting in sudden, extreme price changes on occasion.

In the normal run of things, when you have a large number of investors in a market, and each one is following his own knowledge, intuition and projections, independent of one another, investment decisions by one trader will not affect the market substantially. The wide range of attitudes amongst investors assures that each fluctuation will be met by a wide variety of responses. Investors will respond to exogenous events and therefore the market may go up or down, but it will tend to do so in an orderly manner. Markets may go through significant gyrations in response to major events such assassinations, war, etc., but the response will be tempered by the variety of responses by investors.

However, when (a) a financial market includes a number of large investors with great influence on supply and demand for securities, (b) these large investors have an essentially identical strategy for investing, and (c) that strategy involves buying during up-trends and selling during downtrends (aka *portfolio insurance*), the stage is set for very exaggerated price movements of securities because any trend (up or down) tends to get amplified by these investors. Consider for example, a mild downtrend that might get started due to a random fluctuation, or more likely, some external event in the world. The large investors are pre-programmed (sometimes with automatic computer-generated sell orders) to sell securities in this downtrend. As their securities are offered up for sale, the market is flooded with sell orders and the prices of securities drop further. This, in turn, causes further selling, producing further price drops, etc. Now, even those smaller investors who are not pre-programmed begin to panic and sell. Along the way, there may be margin calls for those who have bought on margin, causing further selling. In an era of investment in the momentum (rather than the value) of securities, when the momentum turns sharply negative, a large number of investors want to get out at the same time. This adds to the selling pressure. Danielsson and Shin refer to this phenomenon as "endogenous risk"—it is the risk inherent in a system with positive feedback that can greatly amplify price movements of securities.

A market may also acquire additional amplification from arbitrage operations. In the case of the stock market crash of 1987 (see "The Crash of 1987" in Chap. 2), the contrast between the prices of stocks and stock market futures played a role when futures were cheaper than stocks, causing arbitragers to sell stocks and buy futures, thus driving stocks down more. In the case of Long-Term Capital Management (see "Long-Term Capital Management" in Chap. 2), arbitraging between Japanese and US currencies led to large losses due to wild variations in the currency markets.

Other factors can contribute to positive feedback, amplifying such market trends. For example, automatic stop-loss orders generate additional selling pressure when the market drops precipitously, and covering short positions can drive a market sharply upward during an uptrend.

Danielsson and Shin concluded:

Endogenous uncertainty matters whenever there is the conjunction of (i) traders reacting to market outcomes and (ii) where the traders' actions affect market outcomes. These conditions are most likely to be in effect when there is a prevailing orthodoxy concerning the direction of market outcomes, and where such unanimity leads to similar positions or trading strategies. In such an environment, the uncertainty in the market is generated and modified by the response of individual traders to the unfolding events. Recognizing these features is essential to intelligent risk management that takes account of endogenous risk.

WHEN THE BUBBLE POPS

K&A raised questions regarding whether and how governmental authorities should respond to a bubble or its inevitable demise.

> Should governmental authorities intervene to cope with a crisis, and if so, at what stage? Should they seek to forestall increases in real estate prices and stock prices as the bubble expands so the subsequent crash will be less severe? Should they prick the bubble once it is evident that asset prices are [excessive]? When asset prices begin to fall, should the authorities adopt any measures to dampen the decline and ameliorate the consequences?

The question as to whether authorities can be effective in forestalling increases in real estate prices and stock prices from becoming excessive remains purely academic because there do not seem to be any examples where authorities have done this. On the contrary, it appears that in every instance, central banks would rather foster growth of the bubble, producing temporary euphoria in the public, than risk a loss of public support for the current administration. In fact, the US Federal Reserve System has systematically fueled the growth of bubbles through-out the 1990s and the 2000s by continually increasing the money supply and lowering interest rates at times of overspeculation. This was compounded by the government's view that deregulation of banks was equivalent to no regulation of banks, and bankers speculated excessively with FDIC-backed funds.

The next question is what government action is desirable when the bubble implodes? As K&A discussed, one point of view is that the best remedy for a panic resulting from an imploding bubble is to let it run its course, and to allow the economy to adjust to the decrease in household wealth that follows from the declines in prices of real estate, stocks, and commodities. According to this view, government intervention encourages formation of the next bubble because "many of the market participants will believe that their possible losses will be limited by government measures." Thus the "likelihood and the scope of future losses" are reduced—at least in the minds of speculators. Speculation is encouraged by government intervention. K&A imply that the government's view is negative toward bubble formation and inadvertently promotes bubbles by their intervention (the road to hell is paved with good intentions). K&A refer to "the undeserved reward to the speculators." However, it seems likely that the Federal Reserve does not view bubbles in such a negative manner.

According to K&A, the view that a panic should be allowed to pursue its course has two elements:

> One element takes pleasure in the troubles that the investors or speculators encounter as retribution for their excesses. . ... The other sees panic as a thunderstorm . . . that clears the air.

K&A provided extensive historical illustrations. The opposing view concedes that while it is desirable to purge the system of bubbles and manic investments, there is the risk that a deflationary panic would spread and wipe out sound investments by non-speculators. As K&A asserted, we will never know whether benign neglect is a good path out of an imploding bubble because this is never done. K&A pointed out that the authorities always feel compelled to intervene. For example, in late 2007, US government authorities gradually became aware that the simultaneous bubbles in real estate and stocks of 2002–2007 was unsustainable and perched precariously on a ephemeral foundation of speculative mortgages based on the expectation that housing prices would advance at 10–20% per year forever.[98] From October through December 2007, a series of shocks propagated through the banking system as the extent of the debacle slowly became revealed. With each revelation, the stock market faltered, and with each reverberation in the stock market, the Federal Reserve came riding to the rescue with a rate cut. Initially, investors responded with enthusiasm, and each rate cut brought on a booming buying spree in stocks. However, it gradually became apparent that these rate cuts had more symbolic value than real value, and by 2008, rate cuts no longer seemed to produce quite as much enthusiasm. In the past twenty-five years, the US Federal Reserve System seems to have adopted its major *raison d'etre* as adapting monetary policy to prop up stock and real estate bubbles. Starting with Greenspan, and continuing with "Helicopter Ben" Bernanke, the goal of the Federal Reserve has been to prop up and maintain bubbles, not so much as a matter of belief and philosophy, but rather because the alternative is viewed as being worse.

In addition to monetary remediation of the popped bubble, President George W. Bush and the US Congress adopted a series of fiscal stimuli, including further tax breaks for businesses (i.e. the rich) along with tax rebates of up to $1,200 per family. The cost to the government of this rebate program has been estimated to be around $160,000,000,000, adding significantly to the already large budget deficit. One must wonder, if the government can simply hand out $1,200 to each family, why not $12,000, or $1,200,000? Why doesn't the government simply make everyone rich? The $160,000,000,000 that the government borrowed to pay out these rebates will never be paid back.

In 2008, the Government interceded a number of times to prop up large business enterprises that failed from speculation. This included bailing out Bear-Stearns, taking over Fannie Mae and Freddie Mac, and essentially taking over AIG Group. The cost of these ventures will certainly be trillions of dollars, and since the Government does not have these funds, it will have to borrow them. it should be noted that the Republican administration that has taken over these private enterprises is the same administration that for years has incessantly repeated the mantra that Government can do no good, and we need to leave everything to private industry.

> Thus, we see that governments in general, and the United States in particular, always act to prop up bubbles. There seem to be only two bullets in the Government's arsenal. One is to cut taxes and the other is to lower interest rates and pump money into the banking system. Both of these will drive up the price of oil, fuel inflation, depreciate the dollar, and promote bubble formation. Inevitably, the Government's cure for excessive spending and inadequate revenues is to increase spending and cut revenues.

NOTES

[1] J. K. Galbraith, "A Short History of Financial Euphoria", Penguin Books, 1993.

[2] It was for this reason that the head of the Federal Reserve, Ben Bernanke, was often referred to as "Helicopter Ben" for dropping money down on the banking system and bailing out troubled institutions.

[3] While many economists believe that low capital gains taxes promote business expansion, the data suggest that this expansion seems to occur in the form of a bubble in the price of paper assets—which primarily benefits the rich.

[4] http://www.stateofworkingamerica.org/tabfig/03/SWA06_Table3.1.jpg

[5] http://www.visualizingeconomics.com/2007/11/04/has-middle-americas-wages-stagnated/

[6] Kindleberger, C. P. and Richard Aliber, "Manias, Panics and Crashes", 5th Edition, Wiley, 2005.

[7] Cheap imported goods from China.

[8] J. K. Galbraith, "A Short History of Financial Euphoria", Penguin Books, 1993.

[9] J. K. Galbraith, "The Great Crash—1929", Mariner Books, 1954.

[10] Note that in 2008, the Congress distributed as much as $1,200 to each household in America. The source of these funds is likely to be borrowing from China. China is happy to provide these funds to protect its principal markets, knowing that most of the money will be spent on goods produced in China.

[11] "Work and Wealth for All," http://workforall.net/ineffectiveness_of_monetary_policy_.html

[12] J. K. Galbraith, "The Affluent Society", 40th Anniversary Edition, Mariner Books, 1958–1998.

[13] http://globaleconomicanalysis.blogspot.com/2006/02/inflation-what-heck-is-it.html

[14] http://inflationdata.com

[15] Morris Rosenthal, http://www.fonerbooks.com/cpi-u.htm

[16] "Even the Insured Feel the Strain of Health Costs," Reed Abelson and Milt Freudenheim, New York Times May 4, 2008.

[17] Consumer Price Index Summary, April 16, 2008. http://www.bls.gov/news.release/cpi.nro.htm

[18] J. K. Galbraith, "A Short History of Financial Euphoria", Penguin Books, 1993.

[19] Debt Shock: The Full Story of the World Credit Crisis. By Darrell Delamaide. 280 pp. New York: Doubleday & Co, 1984.

[20] http://query.nytimes.com/gst/fullpage.html?res=9407EED91239F93BA35754C0A9629 48260&n=Top%2FFeatures%2FBooks%2FBook%20Reviews

[21] http://bigpicture.typepad.com/comments/2008/01/are-bankers-inc.html

[22] P. Krugman, "Banks Gone Wild", New York Times, November 23, 2007.

[23] Note that Professor Fisher never suggested that when the markets were leaping upward, they might have been going up because they were going up. And indeed, 75 years later, we still tend to think that bubbling markets are normal and any decline is abnormal.

[24] Ellen R. McGrattan and Edward C. Prescott, "The 1929 Stock Market: Irving Fisher Was Right", Federal Reserve Bank of Minneapolis, Research Department Staff Report 294, December 2003.

[25] J. K. Galbraith, "The Great Crash", Mariner Books, 1954, Chapter II: "Something Should be Done?"

[26] White, Eugene (1990a), "When the Ticker Ran Late: The Stock Market Boom and Crash of 1929," in Eugene White, ed., Crises and Panics: Lessons of History (Homewood, N.J.: Dow Jones-Irwin); White, Eugene (1990b), "The Stock Market Boom and Crash of 1929 Revisited," Journal of Economic Perspectives 4:2, pp. 67–83.

[27] J. Bradford De Long and Andrei Shleifer, "The Bubble of 1929: Evidence from Closed-End Funds," 1990. www.j-bradford-delong.net/pdf_files/Bubble_1929.pdf

[28] "The Causes of the 1929 Stock Market Crash—A Speculative Orgy or a New Era?" Harold Bierman, Greenwood, 1998.

[29] This statement is very revealing. Bierman made the valid point that one may compare the speculative excess of the dot.com craze in the late 1990s to that of 1929. In fact, this is displayed in Figure 1.3. However, Bierman, under the apparent belief that the dot.com bubble was legitimate and would not subsequently crash, suggested that the bubble of 1929 was no worse (or at least not much worse) than that of the late 1990s—which seems like a reasonable claim. Hence the legitimacy of the dot.com bubble would convey legitimacy upon the 1929 bubble. However, writing in 1998, he had no inkling that the dot.com bubble would burst in 2000, a year and a half later, thus reversing his argument, and leading to the conclusion that the 1929 bubble was just as excessive as the 1998 bubble.

[30] "The Great Bull Markets 1924–1929 and 1982–1987: Speculative Bubbles or Economic Fundamentals?" G. J. Santoni, Senior Economist, Federal Reserve Bank of St. Louis, November, 1987.

[31] In doing this, Mr. Santoni clearly implied that these claims are specious and overly emotional.

[32] I am reminded of an occurrence in my life in 1954. Having completed a course in physical chemistry, and attending a class in chemical engineering in college, the professor asked the class: "I have a vat of liquid that is being heated. How do I know when it is boiling?" One student said: "When the vapor pressure equals the atmospheric pressure." No good. Another student said: "When the vapor pressure of the liquid equals the sum of atmospheric pressure plus liquid head." No good. The professor said: "You know it is boiling when bubbles of vapor form in the liquid and rise up to the surface." As students, we were so imbedded in mathematics that we lost sight of the physical reality.

[33] Paul R. LaMonica, "It's Inflation Stupid," April 15, 2008, CNN.Money.com

[34] http://www.actionforex.com/fundamental-analysis/fed/(fed)-richard-w.-fisher-%11-challenges-for-monetary-policy-in-a-globalized-economy-2008011735172/

[35] "A Brief History of the 1987 Stock Market Crash with a Discussion of the Federal Reserve Response", Mark Carlson, Staff Working Paper 2007-13, http://www.federalreserve.gov/pubs/FEDS/2007/200713/200713pap.pdf

[36] "MOVE OVER, ADAM SMITH: The Visible Hand of Uncle Sam", John Embry and Andrew Hepburn, Sprott Asset Management Special Report, August 20, 2005 http://www.sprott.com

[37] Since stocks are owned predominantly by the rich, this implies that preservation of the wealth of the rich is integral to American preeminence and world stability.

[38] "Monetary Policy Should Gently Lean against Bubbles" in "Irrational Exuberance", Robert J. Shiller, Doubleday, 2nd ed., 2004.

[39] J. K. Galbraith, "The Economics of Innocent Fraud", Houghton-Mifflin, 2004.

[40] The Republicans seem particularly adept at using slogans to lure simple-minded people into their camp. For example, they have gotten poor people to vote Republican

by referring to Democrats as "tax and spend" even though the Democrats represent the poor much more than the Republicans. Similarly, the Republicans have justified continuance of an illegal, unjust, expensive war in Iraq by referring to the opposition as "cut and run."

[41] The Center on Budget and Policy Priorities, The State of the Estate Tax as of 2006, by Joel Friedman and Aviva Aron-Dine, http://www.cbpp.org/5-31-06tax2.htm

[42] I have searched far and wide for an explanation of why the 2007 exemption was scheduled to drop this much, but I cannot find any.

[43] http://mwhodges.home.att.net/nat-debt/debt-nat-a.htm

[44] http://www.cbpp.org/policy-points4-18-08.htm

[45] A recent publication is: Edward N. Wolff, "Recent Trends in Household Wealth in the United States", Working Paper No. 502, The Levy Economics Institute of Bard College, June, 2007.

[46] By transferring funds from the middle class to the poor, the rich open up new markets for their products. I am indebted to my friend Giulio Varsi for pointing this out.

[47] David Cay Johnston, "Perfectly Legal", Portfolio, Penguin Books, 2003.

[48] Shiller, Robert J., "Irrational Exuberance", 2nd Edition, Doubleday/Random House, 2005.

[49] J. K. Galbraith, "The Affluent Society", 40th Anniversary Edition, Mariner Books, 1958–1998.

[50] The Internet is full of websites that justify the wealth of the rich.

[51] In fact they have committed murder to prevent abortion.

[52] However, this seems to merely say that the reason inequality has faded is because it has faded.

[53] Debt Shock: The Full Story of the World Credit Crisis. By Darrell Delamaide. 280 pp. New York: Doubleday & Co, 1984.

[54] A recent publication is: Edward N. Wolff, "Recent Trends in Household Wealth in the United States", Working Paper No. 502, The Levy Economics Institute of Bard College, June, 2007.

[55] However, this assumes "normal" times when banks will only loan up to ~80% of the appraised value of a house. During the real estate bubble of 2002–2007 in the United States, loans were made for more than 100% of the (inflated) appraised value making these loans very risky.

[56] In "normal" times, real estate values tend to be relatively stable. However, during the real estate bubble of 2002–2007 in the United States, and the ensuing aftermath when the bubble popped in late 2007, real estate values went through significant variations.

[57] http://www.cedarcomm.com/~stevelm1/usdebt.htm#_ftn4

[58] http://mwhodges.home.att.net/nat-debt/debt-nat-a.htm

[59] I say "apparently" because it is difficult to be sure what this is.

[60] The Social Security Network, A Century Foundation Project, http://www.socsec.org/publications.asp?pubid=540

[61] United States National Debt (1938 to Present)—An Analysis of the Presidents Who Are Responsible for the Borrowing, By Steve McGourty, http://www.cedarcomm.com/~stevelm1/usdebt.htm

[62] Debt Shock: The Full Story of the World Credit Crisis. By Darrell Delamaide. 280 pp. New York: Doubleday & Co, 1984.

[63] Los Angeles Times, July 24, 2008.

[64] Jospeph Stiglitz and Linda Bilmes, "The True Cost of the Iraq War," W. W. Norton Co., 2008.

[65] Fannie Mae, 2007.

[66] "Triple-A Failure," Roger Lowenstein, New York Times Magazine Section, April 27, 2008.

[67] Fannie Mae, 2007.

[68] Federal Reserve Statistics Release, March 6, 2008.http://www.federalreserve.gov/releases/Z1/Current/

[69] Jonathan McCarthy, "Debt, Delinquencies, and Consumer Spending," Federal Reserve Bank of New York, Current Issues, February 1997.

[70] Charles Steindel, "How Worrisome Is a Negative Saving Rate?" Federal Reserve Bank of New York, Current Issues, May 2007.

[71] "The Dollar and the Current Account Deficit: How Much Should We Worry?", Michael Mussa, http://www.aeaweb.org/annual_mtg_papers/2007/0106_1015_0704.pdf

[72] Since this was written in January 2007, the dollar had already depreciated a good deal to $1.57 per euro, but recovered to about $1.30 at the end of 2008.

[73] This will reduce imports relative to exports. However, a large proportion of goods purchased in the US are made in China so it is difficult to see how this can be accomplished, except via a recession in the US so as to cut demand.

[74] The sharp rise in petroleum prices in early 2008 has stimulated some tendencies toward more fuel-efficient vehicles in the United States, but it is not yet clear how permanent this will be.

[75] Consumer Bankruptcy and Household Debt by Mark Jickling, Congressional Research Service, Library of Congress, http://209.85.165.104/search?q=cache:OFYnon68 BIsJ:digital.library.unt.edu/govdocs/crs/permalink/meta-crs-2824:1+revolving+consumer+debt+in+the+1990s&hl=en&ct=clnk&cd=5&gl=us&lr=lang_en; also US Courts, http://www.uscourts.gov/Press_Releases/bankruptcyfilings081607.html

[76] American Bankruptcy Institute; and Total Bankruptcy, http://www.totalbankruptcy.com/bankruptcy_law_updates_year_later.htm

[77] Debt Shock: The Full Story of the World Credit Crisis. By Darrell Delamaide. 280 pp. New York: Doubleday & Co, 1984.

[78] http://www.fdic.gov/about/learn/symbol/index.html. The limit was raised to $250,000 in 2008.

[79] "A Brief History of Deposit Insurance in the United States" prepared by the FDIC, 1998.

[80] John Kenneth Galbraith, "The Culture of Contentment." (Houghton Mifflin, 1992).

[81] Debt Shock: The Full Story of the World Credit Crisis. By Darrell Delamaide. 280 pp. New York: Doubleday & Co, 1984.

[82] http://en.wikipedia.org/wiki/Reagan_administration_scandals

[83] John Kenneth Galbraith, "The Great Crash" Mariner Books, 1954.

[84] http://www.reuters.com/article/pressRelease/idUS133442+19-May-2008+PRN20080519

[85] Maggie Mahar, "Bull"—A History of the Boom, 1982–1999, Harper Business, 2003.

[86] The author was an employee of Caltech and a participant in MBLI in 1991. He was one of the lucky few that extricated their funds from MBLI before it went into bankruptcy. Had he suffered with the more than 300 others who were stuck in MBLI, he might not have been able to retire when he did.

[87] http://www.nytimes.com/2007/12/19/business/19pension.html

[88] http://www.state.nj.us/sos2008/speech.html

[89] http://www.ocregister.com/ocr/sections/news/focus_in_depth/article_495528.php

[90] "A History of the Global Stock Market from Ancient Rome to Silicon Valley", B. Mark Smith, University of Chicago Press, 2004.

[91] According to Smith, "a flood of pension and retirement money into equities . . . helped power the great bull markets of the 1980s and the 1990s."

[92] In the film: "Green Card," Gerard Depradieu foists himself off as a musician, and when asked to play, he performs an incredible mish-mash of random notes and clashing chords. When he is done, the audience says: "Wow, what was that?" He replies: "It ain't Mozart."

[93] B. Mark Smith, loc cit.

[94] Maggie Mahar, "Bull"—A History of the Boom, 1982–1999, Harper Business, 2003, p. 13, 21.

[95] It dropped to $2/share in March 2008.

[96] Not to worry about inflation: The Government reported that inflation slowed to zero in February, 2008, just as gasoline prices were sidling up toward $4 a gallon.

[97] "Endogenous Risk", Jon Danielsson and Hyun Song Shin, April 21, 2002, http://hyunsongshin.org/www/risk1.pdf

[98] It should be noted that with a small down payment, a house purchase is highly leveraged, and a 10–20% increase in house price can translate into a much larger percentage profit gain on investment.

Chapter 2

A SHORT HISTORY OF BOOMS, BUBBLES, AND BUSTS

*T*he first documented major boom-bubble-bust cycle was the Holland tulip craze that we already described in the Introduction to this book. The next section of this book presents a review of a number of the important boom-bubble-bust cycles that followed from the 18th through the 21st centuries. We begin with two spectacular cycles from the 18th century, and then move on to the United States in the 20th century.

In "The Great Crash" JKG briefly recounted some of the history of bubbles and crashes in the United States:

> In the United States in the nineteenth century, there was a speculative splurge every twenty or thirty years. This was already a tradition, for the colonies ... had experimented at no slight cost with currency issues that had no visible backing. They did well until it was observed that there was nothing there.
>
> The American Revolution was paid for with Continental notes, giving permanence to the phrase 'not worth a Continental'. In the years following the war of 1812–1814, there was a major real estate boom; in the 1830s came wild speculation in canal and turnpike investment.... This came powerfully to an end in 1837. In the 1850s came another boom and collapse, in which a New England bank closed down with $500,000 in notes outstanding and assets to cover them of $86.48. After the Civil War came the railroad boom and a particularly painful collapse in 1873. Another boom came to an equally dramatic end in 1907.

These are only a few of the many cycles of excess in investing in the United States prior to the modern era. Bordo[1] and Wood[2] provide histories of booms and busts in the United States and the United Kingdom. They documented more

D. Rapp, *Bubbles, Booms, and Busts*, DOI 10.1007/978-0-387-87630-6_2,
© Springer Science+Business Media, LLC 2009

than 20 crashes in the past two centuries. An IMF Report[3] identified 13 stock market bubble-crash sequences from 1800 to 1940, with peak-to-trough drops ranging from 16.4 to 66.5%.

The rich tradition of booms and busts, established early in the history of the United States, was further enlarged in the 20th century, as will be discussed in the ensuing sections.

Then, we provide a brief review of a few recent Asian bubbles. Finally, we provide an insight into what the next bubble might be.

THE NEW WORLD

In the early 1700s, an intrepid entrepreneur (John Law) developed the foundations of modern bubbles with two promotions, the South Seas venture in England and the Mississippi Company in France. Like most bubbles that followed over the next three hundred years, there was actually a basis and a rationale for believing that a great new opportunity was at hand. The opening up and settlement of the New World seemed to offer a vast source of raw materials and products, as well as a large potential market for European products. This was not entirely unlike the advent of the automobile and widespread electrification in the 1920s or the introduction of the Internet in the 1990s. Bubbles are usually based initially on seemingly sound and rational future prospects. Where bubbles often go wrong is in assuming that these prospects can be easily tapped in the immediacy of time. As enthusiasm builds, investors lose sight of the realities of the prospect and focus only on trading paper for profit. John Law was a great financial innovator. He was the first to espouse the use of large-scale credit and printed money as a replacement for hard currency. He was the original Flim-Flam man.[4] If he were alive today, he would likely be made director of the US Federal Reserve System. Many of the corporate manipulators of our time (Milken, Keating, Lay, Rigas, ...) would have been proud to have known John Law and would have paid great homage to him as the founder of their profession.

SOUTH SEAS BUBBLE

In late 1719, John Law circulated his treatise on economics entitled *Money and Trade Considered*. As Smith discussed, the two central ideas in this work were: (1) credit, when circulated, acts as if it were conventional currency, and (2) commercial activity is stimulated by the money supply.[5]

Mr. Law applied these theories to the South Sea Company in Great Britain. The South Sea Company was created to "take over" responsibility for British government debt in exchange for the exclusive right to engage in trade with Spanish America. As would be the case with the subsequent Mississippi Company, holders of government bonds, which traded at significant discounts due to the precarious financial condition of the government, could use their depreciated bonds at face value to acquire South Sea shares. The company agreed to accept reduced interest payments from the government on bonds it received in exchange for its newly issued shares. As in the case of the subsequent Mississippi Company, everyone seemed to benefit; "the government was able to reduce the interest payments on the public debt, bondholders would receive [what seemed to be] full value for their bonds, and the new company itself would presumably be able to reap large profits from future trade with the Americas."

The press was enthusiastic. Wild claims were made for the demand for luxury merchandise in the New World, and "frequent references were made to large deposits of precious minerals in South America." It was widely assumed that the South Sea Company could achieve great profitability from trading with Spanish colonies in the New World.

However, as Smith observed:

> Some observers were skeptical, noting that at the time the South Sea Company was formed, the Spanish government forbade foreign nationals from trading with its colonies, a policy that seemed unlikely to change.

In its early years, the South Sea Company did not do well; the only concession the Spanish allowed was the right to engage in the slave trade. But in 1720, the company engaged in a number of practices that started a speculative binge. It spread extravagant rumors of the value of its potential trade in the New World. It loaned shares to highly placed officials in the government, and bought them back when share prices rose, generating profit for politicians based on no investment. Meanwhile, the South Sea Company acquired an aura of legitimacy on the claim that all the top government officials had "invested" in the company. One website claims that 462 members of the House of Commons and 112 Peers were involved with the company, and that King George I and his two mistresses "were heavily involved in the South Sea Company." The stock price increased eight-fold from January to June 1720. As Wikipedia said:

> Its success caused a country-wide frenzy as citizens of all stripes – from peasants to lords – developed a feverish interest in investing; in South Seas primarily, but

in stocks generally. Among the many companies, more or less legitimate, to go public in 1720 is – famously – one that advertised itself as 'a company for carrying out an undertaking of great advantage, but nobody to know what it is'.

A number of other joint-stock companies then joined the market, making usually fraudulent claims about other foreign ventures or bizarre schemes, and were nicknamed 'bubbles'.

The South Seas Company held a charter providing exclusive access to all of Middle and South America. However, the areas in question were Spanish colonies, and Great Britain was then at war with Spain. Even once a peace treaty had been signed, relations between the two countries were not good. The ... South Sea Company was able to obtain [the right] to send only one ship per year to Spain's American Colonies.

The bubble popped at the end of the summer of 1720, and the stock price dropped by a factor of eight in September. A number of people around the country lost all their money and "the gullible mob whose innate greed had lain behind this mass hysteria for wealth, demanded vengeance. The South Sea Company Directors were arrested and their estates forfeited."

JOHN LAW'S MISSISSIPPI COMPANY

The originator of the South Seas bubble in England got into trouble by killing his opponent in a duel and had to escape to France, where he promptly began a similar scheme called the "Mississippi Company." Smith[6] provided an excellent description of John Law's Mississippi Company.

John Law founded the Mississippi Company and acquired from the French Government the exclusive right to trade with the French Colony of Louisiana. This was coupled to a plan to reduce the French Government's payments on its debt.

John Law used the same basic strategy for the Mississippi Company as for the South Seas Company. The state of the French Government's economy was very poor and French government bonds traded at a large discount. Law worked out a deal with the French Government whereby his Mississippi Company would offer to trade shares in the company to the public for French Government bonds at par, and he would agree to accept lower interest payments from the Government on the bonds he acquired. As in the case of the South Seas Company, this seemed to benefit the French Government, the bondholders and the Company. However, as we perceive in retrospect, the bondholders were trading discounted paper with tangible value for par value paper backed only by dreams and speculations.

As in the South Seas Company, initial public response was lukewarm, so he added a number of new features, such as exclusive rights to "raise tobacco (which was rapidly becoming popular in France) as well as the right to trade in slaves and other products from the French colony of Senegal." All of this was paid for with newly issued shares:

> In order to sell the new stock, Law aggressively hyped the company's prospects in a promotional blitz that resembled a modern public relations campaign.

He continued a wide range of financial schemes and the stock rose by more than a factor of ten in a buying frenzy of the public.

Smith described the buying frenzy of luxury items that resulted from this early stock bubble that produced a "stock market-induced wealth effect – a change in personal consumption patterns arising from dramatic moves in stock prices."

During the same time period that the South Seas bubble expanded and popped, the Mississippi Company went through the same type of cycle – a runaway bubble followed by a popped bubble. John Law's involvement with the French Government and his rampant generation of credit (he even lent government money to people to purchase company shares) led to a severe inflation.

FLORIDA LAND BOOM OF THE 1920s

THE RISE

In 1931, Fredrick Lewis Allen[7] wrote an oft-quoted classic book: "Only Yesterday" that described the era of the 1920s with great insight and perception. One of the major financial events of the 1920s was the Florida land boom, and Allen discussed this in detail. As Allen described it, the boom was building for several years, but reached a frenzy by 1925, when:

> Miami had become one frenzied real-estate exchange with 2,000 real-estate offices and 25,000 agents marketing house-lots or acreage.... The city fathers had been forced to pass an ordinance forbidding the sale of property in the street, or even the showing of a map, to prevent inordinate traffic congestion.

People flooded into Florida to buy and sell land:

> Hotels were overcrowded. People were sleeping wherever they could lay their
> heads, in station waiting rooms or in automobiles. The railroads had been forced
> to place an embargo on imperishable freight in order to avert the danger of
> famine; building materials were now being imported by water and the harbor
> bristled with shipping. Fresh vegetables were a rarity, the public utilities of the
> city were trying desperately to meet the suddenly multiplied demand for elec-
> tricity and gas and telephone service, and there were recurrent shortages of ice.
>
> By 1925 they were buying anything, anywhere, so long as it was in Florida.
> One had only to announce a new development, be it honest or fraudulent, be it
> on the Atlantic Ocean or deep in the wasteland of the interior, to set people
> scrambling for house lots. . . . The stories of prodigious profits made in Florida
> land were sufficient bait.[8]

Allen provided many examples of huge increases in the prices of lots. These stories
were multiplied and spread, adding fuel to the fire. The standard joke at the time was:
"a native saying to a visitor, 'want to buy a lot?' and the visitor at once replied: 'Sold'."

Lots were bought from blueprints. Subdivisions were drawn up, and advertisements described them. "Binders" were made with a check for 10% down payment.[9]

Plans were laid out for new hotels, apartment houses and casinos. Allen
described a sight at the height of the boom where a large vacant lot was almost
completely covered with bathtubs in crates that had been there for some time. The
tubs were intended for apartment buildings but the freight embargo had held up
the remainder of the contractor's building material and after the bathtubs arrived.

Allen also described the advertisements of the time as resounding with "slogans
and hyperboles of boundless confidence." The Miami Daily News printed an issue
of 504 pages (mainly advertisements) one day in the summer of 1925.

By the height of the 1920s land boom, a single piece of land was changing hands
as many as six times a day. "Binder Boys" sold land for a small down payment, the
understanding being that the land would probably sell at a higher price before the
next payment came due. There always seemed to be another buyer hoping to jump
into the market, causing the prices to skyrocket further.

THE FALL

The Florida land boom began to collapse in the spring and summer of 1926. People
who held binders were defaulting on their payments. Many of those with paper profits

found that the properties they owned were preceded by a series of purchases and sales, all at 10% down, and as many of these defaulted, the only options were to either hold onto the land at a great loss, or default. The land was often burdened with taxes and assessments that amounted to more than the cash received for it, and much of the land was blighted with a partly constructed development. As the deflation expanded, two hurricanes added the finishing touch to the bursting bubble. The hurricanes left four hundred dead, sixty-three hundred injured, and fifty thousand homeless.

According to a source quoted by Allen, by 1927, the approach to Miami by road was littered with dead subdivisions:

> ... their pompous names half-obliterated on crumbling stucco gates. Lonely white-way lights stand guard over miles of cement sidewalks, where grass and palmetto take the place of homes that were to be. . . . Whole sections of outlying subdivisions are composed of unoccupied houses, past which one speeds on broad thoroughfares as if traversing a city in the grip of death.

Bank clearings for Miami had climbed sensationally to over a billion dollars in 1925 but dropped sharply after that (see Table 2.1).

Table 2.1 Bank clearings for Miami

1925	$ 1,066,528,000
1926	$ 632,867,000
1927	$ 260,039,000
1928	$ 143,364,000
1929	$ 142,316,000

As Allen summarized:

> Most of the millions piled up in paper profits had melted away, many of the millions sunk in developments had been sunk for good and all, the vast inverted pyramid of credit had toppled to earth, and the lesson of the economic falsity of a scheme of land values based upon grandiose plans, preposterous expectations, and hot air had been taught in a long agony of deflation.

UNDERLYING CAUSES

Allen provided seven contributing factors to the Florida land boom:

1. Florida's climate.

2. Accessibility to the populous cities of the Northeast.

3. Portability of people with automobiles.

4. Aura of confidence pervading the population during the 1920s.

5. The desire to live in a country club environment.

6. The motivation to emulate the success of selling Southern California.

7. The belief that Florida land offered the best chance to get rich quick.

These were all factors that made Florida attractive. In the early 1920s, Florida became a popular place for vacations or relocation because of its climate. The population grew steadily and housing couldn't match the demand, causing prices to increase sharply, which was not exactly unjustified at that point. But, as prices doubled and tripled, the word spread and speculation began. Soon, nearly everyone in Florida was either a real estate investor or a real estate agent. This was a classic case of the phases of speculation as discussed in "The Rationality of Investors" in Chap. 1. Initially, people invested in Florida real estate because it was an attractive location. This caused prices of real estate to rise. In the second stage, as the prices of real estate increased even more, speculators moved in to buy real estate, not to dwell in the housing they own, but with the intent of turning over their holdings to another speculator who would arrive on the scene later, having noted the expanding bubble in housing. In the speculative stage, the original reason for investing in Florida real estate was forgotten, and investments were made only to soon turn over the investment to "a bigger fool." As the frenzy built, speculators borrowed to increase their leverage, and thus expanded the bubble until it eventually popped.

THE ROARING 20s STOCK MARKET

THE REAL ECONOMIC BOOM OF THE 1920s

If you query "Google" on the Internet, you find a huge number of articles and websites that address the stock market crash of 1929 and the ensuing depression of the 1930s, but very few sites that deal with the actual boom of the 1920s, which was clearly a proximate cause of the crash and the ensuing economic depression. The reason for this seems to be that we have a deep and pervasive belief in our society that it is only right and natural that stocks should go up, even by huge percentages, and such increases in asset prices are neither unreasonable nor demanding of explanation. However, when stocks go down, that is considered to be remarkable and deserving of study and examination. When stocks crash, it is a national calamity requiring investigations, allegations and accusations.

There are many factors that contributed to the economic boom of the 1920s.[10] The First World War had accelerated the gradual transition of America from an agrarian nation toward an industrial nation, although agriculture still played a much larger share than it does today. The advent of mass production in electrically powered factories with assembly lines produced products efficiently at low prices. Automobiles became commonplace and the majority of American households owned cars by the end of the 1920s. The automobile and trains revolutionized transportation. The workweek dropped from 60 to 48 h, and Americans had more time for leisure. The consumer outlook was optimistic. Taxes were low and businesses and individuals were able to retain much of their earnings. As Figure 2.4 shows, the uppermost income tax bracket during the second half of the 1920s was 25% – the lowest it has ever been. Similarly, the capital gains tax was the lowest it has ever been. There was an ample money supply. Advertising became a big business and America became a consumer society driven by the urgings of the advertisers.

Unlike the era of the late 20th century, where "free trade" has been widespread among nations, high tariffs were placed on imported goods in the 1920s, promoting production and distribution of goods made in America. This also contributed to American prosperity. However, other nations were not so fortunate. Since the United States relied on its own reserves of national resources:

> ... little money had to leave the United States to buy the raw materials needed to manufacture its products. This created an unbalanced cash flow from the rest of the world to the United States. As a result, European nations, still recovering from the war, needed loans, which they got from American banks. This sent even more money to the United States in the form of repayments and interest, leading to an even more unbalanced cash flow, and so on.[11]

Furthermore, Europe's recovery from World War I did not revive to pre-war levels of production, and the Europeans failed to reclaim their old markets from the United States or create new markets to compensate for the losses. As a result, nations still maintained high tariffs, which raised prices and cut world trade.

The question has been debated by economists for over 100 years, as to whether tariffs promote or oppose prosperity. Most economists in the 21st century seem to be enthusiastically in favor of free trade, and it is widely believed that tariffs contributed to the worldwide depression of the 1930s. There seems to be some merit to this argument. However, like most economic questions, the issues are complex. In an ideal world composed of nations of roughly equal size and gross national product, without cartels and other artificial controls of supply and demand, one may argue that with free trade, each nation can produce the products for which it is most capable and efficient. Without trade barriers, all nations benefit from the most efficient production wherever it occurs, and prosperity is shared by all. Free trade is believed to be a boon in such a world. However, this seems to be the hypothetical world of economists.[12] In the real world of the early 21st century, with the United States being a major industrialized power, and many developing countries anxious to industrialize with cheap labor, free trade may provide short-term advantages to both the United States and the developing nations. The United States allows its manufacturing base to be usurped by the developing nations, and in return the citizens of the United States can purchase cheap goods made by underpaid labor in the developing countries. Eventually, however, the loss of manufacturing in the United States will take its toll with a loss of jobs and independence, and as wages inevitably rise in developing countries, the benefits to the United States will gradually disappear, and the United States will be worse off. While the United States has been actively pursuing this policy from about 1990 to 2008, the days of reckoning seem to be approaching. Furthermore, in a world where oil plays a dominant role, and oil resources are distributed sporadically amongst mainly reactionary and often despotic nations, there cannot be any actual "free trade." Eliot Janeway (*The Economics of Chaos*) said:

> Americans rushed to buy import bargains, even while suspecting how much their savings as shoppers would cost them income earners. Dogmatism rooted in the cliches of free trade hypnotized the victims into welcoming the losses as gains. Optimism fed on the euphoric lure of America's presume growth into a "service" economy free from the import threat-until depression struck the entire service industry, from restaurants to hospitals and even television networks. It jolted the country into learning a basic lesson her economists had never taught her: service industries are intertwined with manufacturing industries. Each relies on the other as a customer. Services cannot continue to enjoy expansion when the manufacturing industries, which produce income to be spent on services, suffer shrinkage. Realism made short shrift of the stubborn rationalization that America could import prosperity.

Reagan swallowed the free-trade dogma, and the country choked. America was first dazzled by the import profiteering that always tops an inflationary boom, and then demoralized by the import dumping that, just as predictably, always leads a deflationary debacle. The import inflation that paced the U.S. sellers' markets of the 1970s collapsed into the import deflation that devastated them in the 1980s. Inescapably, however, the dumping countries suffered along with their target. Their economies started to contract despite the expanded outlet for their goods America was inviting them to buy with subsidized prices and credit. Though America was importing distress from every point on the map by the mid-1980s, her competitors underselling her in her own markets were scarcely exporting themselves into prosperity.

One industry that did not participate in the boom of the 1920s was farming. Farms had expanded greatly during World War I to feed the allies, but European agriculture recovered by 1924–1925, resulting in US overcapacity, leading to widespread misery for US farmers. Over 600,000 US farmers went bankrupt within 5 years.

The boom of the 1920s was initially built on a solid basis. The advent of widespread electrification, the lowered cost and greater distribution of automobiles, and the technical advances in industrialization, all produced an environment conducive to increased economic prosperity. JKG presented data that show that there were real economic gains in the period 1925–1929 (e.g. the value of output rose 13% in five years, the increase in automobile production was 23% in 3 years, and industrial production increased by 64% in 7 seven years (after the down year of 1921)). The Federal Reserve index of industrial production rose from a depressed value of 67 in the recession of 1921 to 100 in 1924 to 126 in mid-1929. Automobile production reached 5.4 million vehicles in 1929, an increase of a million over 1926. Wages were not going up much but prices were stable. However, as in all booms, the price of assets soon rose far higher than the real increase in prosperity, producing a bubble in which stock valuations were bid up to very high levels.

THE STOCK MARKET OF THE 1920s

As JKG pointed out,[13] it is difficult to say when the stock market boom of the 1920s began. In the second half of 1924, the New York Times index of 25 industrial stocks (NYTI) rose from about 106 to 134 – a 27% gain in six months. A year later, at the end of 1925, it had increased to 181 for a yearly gain of 35%. These two years provided the formative stage for the bull market of the second half of the 1920s. The next year, 1926 was an off year in which the NYTI lost most of the gains of 1925, reaching as low as 143 in March 1926. However, beginning in 1927, the stock market began advancing, and except for brief temporary setbacks, continued to rise through the summer of 1929. The NYTI proceeded to reach 245 at the end of 1927, 332 at the end of 1928, and 449 in August of 1929.

The Federal Reserve cut the rediscount rate from 4 to 3.5% in 1927, and it is widely believed that this was a major contributor to the further expansion of the stock bubble. However, JKG disparages this belief as being too simplistic. This small cut in a key interest rate could not have opened the spigots of money flow inordinately, and indeed later in the boom, the cost of borrowing became a trivial matter compared to expectations of future gain from securities, when investors were happy to pay double-digit interest rates to borrow funds to invest.

According to JKG, the gains in stock prices through 1927 could possibly be rationalized to some extent as reflecting real gains in the economy, although this requires a stretch of the meaning of "rational." Considering that the NYTI at the end of 1927 had more than doubled in three years, such gains appear to be excessive. There never has been, nor will there ever be a doubling of the US economy in three years – by any reasonable yardstick. However, allowing that prior to 1928, perhaps the stock market was merely "exuberant," JKG argued that in 1928 "the nature of the boom changed," and there was a "mass escape into make-believe" and the "speculative orgy started in earnest." JKG pointed out that in this period:

> [The stock market] did not rise by slow, steady steps, but by great vaulting leaps. On occasion it also came down the same way, only to recover and go higher again.

Another indicator of speculative excess was the huge increase in volume of shares traded per day in 1928 and 1929. Wild oscillations, both upward and downward (but predominantly up) took place, suggesting that valuation of stocks had taken on an extremely subjective basis and economic fundamentals no longer influenced stock prices (as if they usually did).

The folklore of the stock market in the late 1920s is replete with stories of waitresses, taxi drivers, barbers and others who overheard discussions by financiers and based on that information, made a good profit in the stock market. Supposedly, almost everyone in America was turned on to stocks and many people checked with their brokers several times a day.[14] However, it seems likely that the number of people who actually owned stocks was about 1.5 million, and the number of people with substantial stock holdings was considerably less than that.

About 40% of the investing public had margin accounts in which they could borrow funds in order to invest a greater amount of money into stocks. In those days, margin was not regulated, and while some brokers limited margin to 50% (i.e. one could borrow up to one dollar for each dollar invested providing 2:1 leverage), some brokers allowed buying stocks on 10% margin (providing 10:1 leverage). For investors buying on 10% margin, their total investment would be wiped out by a 10% drop in stock prices, necessitating a "margin call" by the broker requiring the investor to either put up more cash, or have his stock sold out at the market price.

Believing that margin loans had been a key element of the stock market boom and crash in the late 1920s, the Federal Reserve Bank was empowered to regulate margin lending with the Securities and Exchange Act of 1933. Ricke provided an analysis concluding: "the availability of margin loans can cause the development of a stock market bubble through inducing investors to pay more for a stock than its fundamental value."[15]

In the 1920s, stock prices were typically in the range of a few hundred dollars per share, and since dollars were worth a good deal more then than they are worth today, a round lot of 100 shares represented a significant investment. Smaller investors did not have enough cash to diversify their portfolios. Investment trusts were invented as a means of providing such diversification to investors, who bought shares in a trust, and the trust maintained a diversified portfolio. These trusts were the forerunners of modern mutual funds. The investment trusts were also used to increase leverage. The investment trusts could increase their leverage by issuing bonds to borrow money from the public, and invest those funds in the stock market.

Figure 2.1 shows the Dow-Jones Industrial Average during the 1920s and 1930s. The advance began in the mid-1920s and accelerated in the late 1920s. The stock market began to falter in September 1929, but then quickly recovered. It began to pass though large gyrations, upward and downward – a good sign that the end was near.

During the month of October, 1929, jumps of as much as 10% in one day occurred, and the NYTI, which had peaked at 449, oscillated between 292 and 415,

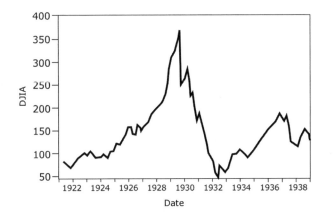

Figure 2.1 Dow-Jones industrial average in the 1920s and 1930s.

ending the month at 344 (down 22% from the August peak). On Tuesday, October 29, 1929, an unprecedented 16,410,030 shares were traded and the market suffered a loss of about 13%. During the week the market lost 30% of its value.

By November 13, the NYTI dropped to 235, a 48% drop from the peak. As one can see in Figure 2.1, the stock market continued to drop precipitously into 1932, when it actually finished below the low point of 1922.

THE CRASH OF 1929

Much has been written about the stock market crash that began in 1929 and the depression of the 1930s – far more than has been written about the spectacular rise of the 1920s. Many explanations have been offered. Most of these lean toward the view that the rise of the 1920s was not acutely abnormal, and special circumstances brought about the downfall of the markets and the ensuing depression. One common "explanation" is based on credit tightening. For example, Shiller[16] claimed that credit tightening was an important contributor to the crash of 1929 and the ensuing depression. He said:

> There have been occasions on which tightened monetary policy was associated with the bursting of stock market bubbles. For example, on February 14, 1929, the Federal Reserve raised the rediscount rate from 5 to 6% for the ostensible purpose of checking speculation. In the early 1930s, the Fed continued the tight monetary policy and saw the initial stock market downturn evolve into the deepest stock market decline ever, and a recession into the most serious US depression ever.

However, JKG argued that nothing could be further from the truth. First of all, the stock market inflated unabated after the February 1929 increase in the discount rate. The stock market simply shrugged off the increase in the interest rate. Second, the bubble mentality was so frothing that investors were happy to pay double-digit margin interest rates[17] to plough more money back into the stock market. According to JKG, what might have contributed more to the demise of the bubble and formation of the depression was fiscal policy in which taxes were raised to balance the budget – which was what JKG calls the "conventional wisdom" of the times.

Some have argued that stocks were not fundamentally overpriced in 1929, and the crash was the result of unfounded public hysteria. As we discussed in "Rationality of Experts," in Chap. 1 the Federal Reserve System published a report in 2003 that concluded:

Even at the 1929 peak, *stocks were undervalued relative to the prediction of theory*.

Apparently, the Federal Reserve believes that even after the spectacular rise in stock prices of the latter half of the 1920s, stocks still remained underpriced. However, as we pointed out at the end of "Monetary Policy and the Federal Reserve System" in Chap. 1, JKG described the Federal Reserve in 1929 as "a body of startling incompetence" and there seems to be no reason to suggest that this has changed in the past 80 years.

What seems to be missing from the explanations for the crash (except for writings of JKG) is that the crash was a natural and unavoidable consequence of the speculative rise that preceded it.

The connection between the stock market crash and the ensuing depression was also discussed at length by many commentators. Did the stock market crash cause the depression, or was the crash merely an indicator of the coming depression that resulted from more profound causes? The crash of the stock market in 1929 is widely believed to have contributed to the ensuing depression, although the exact mechanism by which this connection was made is not clear. We discuss this further in "The Great Depression" in Chap. 2.

THE GREAT DEPRESSION OF THE 1930s

The great depression of the 1930s began subtly in the summer of 1929, and picked up steam into 1930. It lasted about ten years and was unmatched for duration and depth by any other economic depression of our time. It is likely that the advent of World War II was a factor in the ending of the depression.

Samuelson described the great depression:[18]

The Great Depression of the thirties remains the most important economic event in American history. It caused enormous hardship for tens of millions of people and the failure of a large fraction of the nation's banks, businesses, and farms. It transformed national politics by vastly expanding government, which was increasingly expected to stabilize the economy and to prevent suffering. . . . President Franklin Roosevelt's New Deal gave birth to the American version of the welfare state. Social Security, unemployment insurance, and federal family assistance all began in the thirties.

It is hard for those who did not live through it to grasp the full force of the worldwide depression. Between 1930 and 1939 US unemployment averaged

18.2%. The economy's output of goods and services (gross national product) declined 30% between 1929 and 1933 and recovered to the 1929 level only in 1939. Prices of almost everything (farm products, raw materials, industrial goods, stocks) fell dramatically.... World trade shriveled: between 1929 and 1933 it shrank 65 percent in dollar value and 25 percent in unit volume. [The depression was world-wide.] Most nations suffered.

According to Romer,[19]

The fundamental cause of the Great Depression in the United States was a decline in spending (sometimes referred to as aggregate demand), which led to a decline in production as manufacturers and merchandisers noticed an unintended rise in inventories.

That may be true, but what caused the cause? Why did this "decline in spending" take place? The "decline in spending" seems to be a symptom, not a cause.

Romer then expounded on several other factors that influenced the downturn. The crash of the stock market in 1929 is widely believed to have contributed to the ensuing depression, although the exact mechanism by which this connection was made is not clear. According to Romer, the stock market crash "generated considerable uncertainty about future income, which in turn led consumers and firms to put off purchases of durable goods." According to this view, people may not have been much poorer but they felt poorer. With a sharp decline in spending, production fell rapidly in 1929 and 1930. Romer suggested:

While the Great Crash of the stock market and the Great Depression are two quite separate events, the decline in stock prices was one factor causing the decline in production and employment in the United States.

However, as JKG pointed out, even though there was a great euphoria in the 1929 stock market, only a small fraction of Americans actually owned stocks.

Romer also indicated that banking panics and monetary contraction were factors in creating the depression.

Prior to origination of the FDIC, the United States experienced widespread banking panics from 1930 to 1933. By 1933, one-fifth of the banks in existence at the start of 1930 had failed. The panics caused a dramatic rise in the amount of currency people held (in their mattresses?) relative to their bank deposits. This reduced the effective money supply, and contraction by the Federal Reserve added to the problem. According to Bordo, the consensus view by economists is that the

1929 crash had a major impact in producing a recession in 1930. This recession deepened to the Great Depression late in 1930 when the Fed failed to prevent a series of banking panics that erupted in the next three years. "The banking panics in turn impacted the real economy through the collapse in money supply, which produced massive deflation.... The depression spread abroad through the fixed exchange rate links of the classical gold standard."[20]

As JKG pointed out,[21] the "conventional wisdom" was "a set of platitudes that have been repeated incessantly until many people believed them – despite the lack of verification," and "when put to the test, the evolution of events often proves the conventional wisdom wrong." The conventional wisdom held that when the economy steered off course, the ultimate remedy was to balance the federal budget. However, as JKG emphasized, taxes had to be raised to achieve a balanced budget, and "it would be hard to imagine a better design for reducing both the private and the public demand for goods, aggravating deflation, increasing unemployment, and adding to the general suffering." Nevertheless, it was widely believed that a balanced budget was just the thing to deal with the depression.

JKG quoted President Hoover in the early thirties who called the balanced budget an "absolute necessity; the most essential factor to economic recovery; the imperative and immediate step; indispensable; the first necessity of the Nation; and the foundation of all public and private financial stability." According to JKG, the "conventional wisdom" dictated policies that were certain to make matters worse during the depression. For example, "Franklin D. Roosevelt was elected in 1932 with a strong commitment to reduced expenditures and a balanced budget...." In 2008, our "conventional wisdom" is just the opposite; we now believe that the Government should just keep distributing money and borrow, borrow, borrow. Under no circumstances does any politician in 2008 desire to balance the federal budget. That probably represents the collective wisdom of economics over the past century.

Romer discussed the belief of some economists that the Federal Reserve's goal of *preserving the gold standard* for American currency caused huge declines in the American money supply. Had the Federal Reserve expanded the money supply, foreigners could have lost confidence in the United States' commitment to the gold standard, leading to large gold outflows and the United States could have been forced to devalue. The effect of these factors on foreign countries was described by Romer:

The deflation in America made American goods particularly desirable to foreigners, while low income reduced American demand for foreign products. To counteract the resulting tendency toward an American trade surplus and foreign gold outflows, central banks throughout the world raised interest rates.

Maintaining the international gold standard, in essence, required a massive monetary contraction throughout the world to match the one occurring in the United States. The result was a decline in output and prices in countries throughout the world that also nearly matched the downturn in the United States.

Toward the end of the 20th century, most economists were enthusiastically in favor of free trade, and believed that free trade is a necessary ingredient of burgeoning national economies. This has been translated into policy by many governments. During this time period, America allowed most of its manufacturing capability to move to Asia (predominantly China) so it could buy cheap products from them (for a while, until wages rise in Asia – which they will). However, the Chinese demand for oil and other commodities drove those prices up, resulting in severe problems for the US balance of payments in importing oil, with a much weaker dollar. Free trade may prove helpful to the developing nations; it will likely prove to be a disaster for the United States.

In the 1930s, economists were not favorable to free trade. The 1930 enactment of the Smoot-Hawley tariff in the United States was meant to boost farm incomes by reducing foreign competition in agricultural products. But other countries followed suit, both in retaliation and in an attempt to force a correction of trade imbalances. Romer asserted:

Scholars now believe that these policies may have reduced trade somewhat, but were not a significant cause of the Depression in the large industrial producers. Protectionist policies, however, may have contributed to the extreme decline in the world price of raw materials, which caused severe balance-of-payments problems for primary-commodity producing countries in Africa, Asia, and Latin America and led to contractionary policies.

However, some analysts believe that the restrictions on trade contributed significantly to the depression.

Samuelson's view was:

The depression [was] the final chapter of the breakdown of the worldwide economic order. The breakdown started with World War I and ended in the thirties with the collapse of the gold standard. As the depression deepened, governments tried to protect their reserves of gold by keeping interest rates high and credit tight for too long. This had a devastating impact on credit, spending, and prices, and an ordinary business slump became a calamity.

Samuelson listed four major factors contributing to the depression:

1. *The gold standard.* Governments had to maintain gold reserves to back up paper money. This limited their ability to expand the money supply to stimulate the economy. A loss of gold (or convertible currencies), forced governments to raise interest rates which had a depressing effect on the economy. One view is that the great depression was "the last gasp of the gold standard."

2. *Economic policy.* According to Samuelson, "economic policy barely existed. There was little belief that governments could, or should, prevent business slumps." However, JKG pointed out that there was a policy, and that policy was to balance the budget – which in this instance was counterproductive. [Note that by 2008, the pendulum had swung so far that a principal objective of the Federal Reserve is to counter business slumps, even those produced as the aftermath of excessive speculation.]

3. *Production patterns.* Samuelson said:

   Farming and raw materials were much more important parts of the economy than they are today. This meant that lower commodity prices could cripple domestic prosperity and world trade, because price declines destroyed the purchasing power of farmers and other primary producers (including entire nations). In 1929 farming accounted for 23 percent of US employment (versus 2.5 percent in 2008). Two-fifths of world trade was in farm products, another fifth was in other raw materials.

4. *Impact of World War I.* According to Samuelson, wartime inflation, when the gold standard had been suspended, had impacted the stability of international relationships between currencies that raised prices and inspired fears that gold stocks were inadequate to provide backing for enlarged money supplies.

Samuelson further discussed the lack of action by the Federal Reserve in permitting two-fifths of the nation's banks to fail between 1929 and 1933. Since deposits were not insured then, the bank failures wiped out savings and shrank the money supply by 1/3. Friedman and Schwartz[22] argued that it was this drop in the money supply that strangled the economy. They consider the depression originally an American affair that later spread abroad.

The literature on the depression is extensive. There are many theories for the causes of the depression developed by economists. Economists tend to seek technical economic factors, and it is likely that most of these were in some part, contributing factors. The underlying belief in these analyses seems to be that the public is assumed to be rational, and with proper and appropriate economic policies, all will go well in the market economy. However, there is a psychological element to the market economy, and like the psychology of an individual, the psychology of the ensemble of people can rise to euphoric heights and crash to severe depression. It seems likely that in addition to the specific economic policy issues of the time, there was a significant psychological factor in the great depression. Like particles and antiparticles in physics, the antithesis of the expanding bubble driven by greed of the herd can be expressed as a herd mentality driven by fear. Similarly, in seeking explanations for many bubbles after they pop, economists tend to search for technical factors but do not often mention the psychological herd instincts of excessive greed or fear.

THE SAVINGS AND LOAN SCANDAL OF THE 1980s

THE ORIGINAL PROBLEM

In 1980, the FSLIC insured approximately 4,000 state-chartered and federally chartered savings and loan institutions with total assets of $604 billion. The vast majority of these assets were held in traditional S&L mortgage-related investments. Another 590 S&Ls with assets of $12.2 billion were insured by state-sponsored insurance programs in Maryland, Massachusetts, North Carolina, Ohio, and Pennsylvania. One-fifth of the federally insured S&Ls, controlling 27 percent of total assets, were permanent stock associations, while the remaining institutions in the industry were mutually owned.[23]

Two books provide lengthy descriptions of the S&L scandal of the 1980s.[24] In this section, I rely heavily on Lowy's book.

The Savings and Loan (S&L) institutions of America were founded for the purpose of providing funding for residential homes. Prior to the late 1970s, most home mortgages were at a fixed interest rate, typically for 30 years, but occasionally for shorter terms. S&Ls were highly regulated and typical requirements included:

- Fixed upper limits to the interest they were allowed to pay on deposits.

- They were not allowed to borrow long-term.

- Requirements for maintaining capital as a percentage of assets.

- Business was limited to loans for residential housing. Loans were not permitted for non-residential construction, raw land investment, or other enterprises. (However, Texas approved a major liberalization of S&L powers allowing property development loans of up to 50% of net worth starting in 1967).

- Many S&Ls were mutual. For those that were owned by stockholders, a S&L was required to have at least 400 stockholders with no single investor owning more than 25 percent of the stock.

- Typically, mortgages were only issued if the mortgagee paid down a significant down payment (at least 20% of the value of the property) so the Loan-to-Value (LTV) ratio was equal to or less than 0.8.

Texas, California, other states, and the federal government liberalized these constraints in the early 1980s when S&Ls developed financial problems.

Although not widely recognized in the 1970s, and possibly still not widely recognized today, the whole concept of banking institutions providing mortgages depends on the relationship between: (1) short-term interest rates paid by the banks on deposits and certificates of deposit (CDs) to acquire funds, and (2) long-term interest rates on mortgages that produce income for the banks. This process for funding residential housing is dependent on the stability of interest rates, and constraints on inflation. Typically, a S&L must have a spread between average mortgage interest and average depositors' interest of about 2.25% in order to cover operating costs and thus break even. While a S&L may initially lend money on a mortgage at an interest rate greater than 2.25% above the current average of depositors' interest, if general interest rates in the country rise as the years go by, the spread between mortgage interest and average depositors' interest may shrink, and the bank may start to lose money on its loans. As it loses money, it may have to eat into its capital, and eventually become insolvent (when its capital becomes negative). On the other hand, if interest rates remain comparatively stable over the years, the mortgages will continue to generate a net profit for the S&L. Another problem for S&Ls is that when short-term interest rates rise, competitive interest rates from other sources (money market accounts, government notes, ...) tend to siphon off funds from the S&L if the rates are higher than the S&Ls are allowed to pay depositors.

Thus we see that the entire structure of the banking system for residential mortgages is fundamentally unstable because any future rise in short-term interest rates will make the income from long-term mortgages inadequate to cover bank expenses, and the bank will operate at a deficit. This became a major problem at the end of the 1970s and into the 1980s as interest rates soared.

As Lowy said:

The interest rate sensitivity mismatch that devastated the S&L industry in the early 1980s [See Figure 2.2] was built into its basic design. The two roles that society had assigned to S&Ls were to provide long-term credit to homeowners at stable rates and to get the money to lend by taking deposits from individual savers. The savers' money could be withdrawn at any time because that was good for the public; the mortgages had to be fixed-rate and long-term because that also was good for the public. The S&Ls weren't allowed to borrow long, and just about the only investments they were allowed to make were long-term mortgages.

It is noteworthy that in the aftermath of the 2000–2001 dot.com stock market debacle, the Fed pushed down interest rates in a frantic effort to resurrect the stock market bubble that preceded 2000, and in the process, millions of homeowners refinanced their mortgages at these new lower rates.

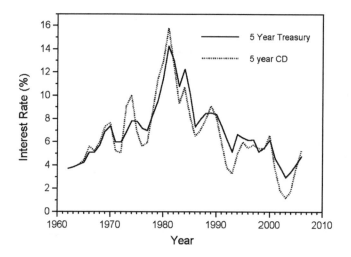

Figure 2.2 Variation of interest rates from 1960 to 2006. (Data from http://www.feder alreserve.gov/RELEASES/h15/data.htm).

> With banks holding long-term mortgages paying low interest rates, if short-term depository interest rates rise in the future – as they likely will, (considering the inflationary policies of Fed monetary policy in 2007–2008) there could be another wholesale failure of the banking system in the future analogous to the S&L problem of the 1980s.

As can be seen from Figure 2.2, interest rates varied over a fairly narrow range in the 1970s until they started rising sharply around 1977. Nevertheless, only about 200 (of 5,000) S&Ls lost money in the 1970s.

As Lowy said:

Amazingly, for many years what was good for the public was good for the S&Ls. . . . [S&Ls benefited] for many years because most of the time long-term rates were higher than short-term rates and interest rates, generally were stable (albeit with an upward bias) from World War II until 1966, when interest rates started to rise more significantly in response to forecasts of inflation.

However, with the sharp rise in interest rates toward the end of the 1970s, things rapidly got worse. In 1980, 1,800 (of 5,000) S&Ls lost money and in 1981 and 1982, 4,000 (of 5,000) S&Ls were in the red.

The seeds of the problem were planted in 1966 when, as Lowy summarized:

In order to protect thrift institutions from having to pay the rising market rates of interest that they couldn't afford, Congress put interest rate controls on savings deposits at all federally insured institutions. No bank or S&L, Congress decreed, could pay more interest on a depositor account than the regulators permitted. In effect, by interest rate regulation, Congress artificially corrected the mismatch by fixing the rate on savings deposits and thereby making them act as if they had locked-in, long-term rates, just like the mortgages owned by thrifts. Thus the policy of rate regulation permitted S&Ls to flourish again without depriving homeowners of long-term, fixed-rate mortgages. Savers, unaware that the system was cheating them, were made to foot the bill.

As Lowy described it:

At the end of the 1970s, the hitherto apparently successful policy of interest rate regulation unraveled. It no longer worked because inflation had driven interest rates so much higher than the 5.5% regulatory ceiling on passbook accounts at thrifts.... By 1978, Treasury bills were paying over 9%.... Many depositors couldn't resist moving their money out of thrift institutions and into these higher-yielding instruments. In response to this threat to thrift institutions' liquidity, in June 1978 the regulators tried the expedient of letting thrifts pay higher interest rates on six-month CDs while retaining the old ceilings on pass-book accounts.... That helped keep money in the thrift institutions, but it cost them a lot. By the end of 1979, over 20 percent of depositors' money was in six-month accounts at rates over 10 percent. On a deposit base of half a trillion dollars, that 20 percent shift from passbook to six-month accounts would cost the S&Ls $5 billion a year – exactly the amount that the industry had made in its best year. The situation was desperate in places like New York and Chicago, where usury laws had kept down the rates that thrifts could charge to borrowers and sophisticated depositors quickly moved their money to the higher-yielding accounts.

The problem got worse at the end of 1979 when the Federal Reserve Board tightened the money supply in an effort to choke off inflationary pressures. Interest rates hit unprecedented highs in 1980–1982 and that brought disastrous consequences for S&Ls. Either they had to pay out much higher interest than they were taking in from mortgages, or they had to sell mortgages at a discount to temporarily solve their cash-flow problems.

Mutual savings banks and S&Ls were losing money because of upwardly spiraling interest rates and asset/liability mismatch. Net S&L income, which totaled $781 million in 1980, fell to negative $4.6 billion and $4.1 billion in 1981 and 1982 (see Table 2.2).[25]

Table 2.2 Solvency of S&Ls in the 1980s (assets, income and reserves in billions of dollars)

Year	No. of S&Ls	Total assets (TA)	Net income	Tangible capital (TC)	TC/ TA (%)	No. insolvent S&Ls	Assets in insolvent S&Ls	FSLIC reserves
1980	3,993	604	0.8	32	5.3	43	0.4	6.5
1981	3,751	640	−4.6	25	4.0	112	28.5	6.2
1982	3,287	686	−4.1	4	0.5	415	220.0	6.3
1983	3,146	814	1.9	4	0.4	515	284.6	6.4
1984	3,136	976	1.0	3	0.3	695	360.2	5.6
1985	3,246	1,068	3.7	8	0.8	705	358.3	4.6
1986	3,220	1,162	0.1	14	1.2	672	343.1	−6.3
1987	3,147	1,249	−7.8	9	0.7	672	353.8	−13.7
1988	2,949	1,349	−13.4	22	1.6	508	297.3	−75.0
1989	2,878	1,252	−17.6	10	0.8	516	290.8	NA

According to the FDIC report:

At year-end 1982 there were still 415 S&Ls, with total assets of $220 billion, that were insolvent based on the book value of their tangible net worth. In fact, tangible net worth for the entire S&L industry was virtually zero, having fallen from 5.3 percent of assets in 1980 to only 0.5 percent of assets in 1982. The National Commission on Financial Institution Reform, Recovery and Enforcement estimated in 1993 that it would have cost the FSLIC approximately $25 billion to close these insolvent institutions in early 1983. Although this is far less than the ultimate cost of the savings and loan crisis – currently estimated at approximately $160 billion – it was nonetheless about four times the $6.3 billion in reserves held by the FSLIC at year-end 1982.

For a variety of reasons, the FHLBB's examination, supervision, and enforcement practices were traditionally weaker than those of the federal banking agencies. Before the 1980s, savings and loan associations had limited powers and relatively few failures, and the FHLBB was a small agency overseeing an industry that performed a type of public service. Moreover, FHLBB examiners were subject, unlike their counterparts at sister agencies, to stringent [federal government] limits on allowable personnel and compensation. It should be noted that the S&L examination process and staff were adequate to supervise the traditional S&L operation, but they were not designed to function in the complex new environment of the 1980s in which the industry had a whole new array of powers. Accordingly, when much of the S&L industry faced insolvency in the early 1980s, the FHLBB's examination force was under-staffed, poorly trained for the new environment, and limited in its responsibilities and resources. Qualified examiners had been hard to hire and hard to retain (a government-wide hiring freeze in 1980–1981 had compounded these problems). The banking agencies generally recruited the highest-quality candidates at all levels because they paid salaries 20 to 30 percent higher than those the FHLBB could offer.

DEREGULATION AND NO REGULATION

In the late 1970s, deregulation as a generic policy was in the air. The trucking and airline industries had already been deregulated. It seemed obvious that regulations on S&Ls should be relaxed, allowing them to use adjustable rate mortgages to protect against future rate rises. Other forms of deregulation were also considered. However, as Lowy said:

If [short-term] interest rates remained high, deregulating the asset side of the balance sheet wouldn't save many S&Ls, it was too late for that.

But governments often need to "show" that they are doing something about a problem even if what they do is ineffectual.[26]

In March 1980, Congress passed the Depository Institutions Deregulation and Monetary Control Act of 1980 (DIDC). It called for a gradual phase-out of deposit interest rate regulation over six years, allowed S&Ls to operate checking accounts, gave federal S&Ls power to invest in consumer loans and expanded their powers to make various kinds of mortgage loans, and declared state usury laws to be null and void. The 1980 Act also increased the deposit insurance ceiling from $40,000 to $100,000.

Although deregulated deposit accounts permitted S&Ls to attract deposits, their earnings did not improve due to continued high interest rates. Deregulation may have made theoretical sense in some ways, but it couldn't prevent S&Ls from losing money in the short run. Some S&L executives contended that interest rate deregulation made the problem worse.

Lowy provided an analysis of S&L balance sheets and concluded:

A broke S&L will continue to lose money forever. . . . The losses will grow in each succeeding year, and the process won't stop even when the S&L runs out of assets. When there are no more assets, the S&L will still have obligations to pay interest to depositors, and those obligations will have to be funded by taking in more deposits and paying old depositors with new depositors' money. Unless the S&L is closed, the losses will, by definition and without fail, grow forever. And this really happened.

In 1981–1982 the savings banks and S&Ls failed because their spreads (between interest earned and interest paid) declined significantly, and in many cases, even became negative. Eventually, these losses caused many institutions to fail.

Lowy asserted that a S&L could not keep losing money forever, were it not for deposit insurance. Without deposit insurance, depositors would withdraw their money when the bank's finances became shaky. With deposit insurance, depositors can keep their funds in a weak bank, so long as it pays a good rate of interest, knowing that the government will bail them out if the bank fails.[27]

According to the FDIC report, most of the insolvencies of S&Ls in the early 1980s were predictable because of the interest rate mismatch:

What followed, however, was a patchwork of misguided policies that set the stage for massive taxpayer losses to come. In hindsight, the government proved singularly ill prepared to deal with the S&L crisis. The primary problem was the lack of real FSLIC resources available to close insolvent S&Ls. In addition, many government officials believed that the insolvencies were only 'on paper,' . . . and would soon be corrected. This line of reasoning complemented the view that as long as an institution had the cash to continue to operate, it should not be closed.

Former Assistant Secretary of the Treasury Roger Mehle took the position that thrifts did not have a serious problem.

Most political, legislative, and regulatory decisions in the early 1980s were imbued with a spirit of deregulation. The prevailing view was that S&Ls should be granted regulatory forbearance until interest rates returned to normal levels, when thrifts would be able to restructure their portfolios with new asset powers.[28]

Perhaps the most far-reaching regulatory change affecting net worth was the liberalization of the accounting rules for supervisory goodwill.

The Bank Board also attempted to attract new capital to the industry, and it did so by liberalizing ownership restrictions for stock-held institutions in April 1982. The elimination of these restrictions, coupled with the relaxed capital require-ments and the ability to acquire an institution by contributing *in-kind capital* (stock, land, or other real estate), invited new owners into the industry. With a minimal amount of capital, an S&L could be owned and operated with a high leverage ratio and in that way could generate a high short-term return on capital. Legislative actions in the early 1980s were designed to aid the S&L industry but in fact increased the eventual cost of the crisis.

HOW MR. REAGAN MADE A BAD PROBLEM WORSE

When the Reagan administration took office on January 20, 1981, Donald Regan's Treasury Department formulated a simplistic set of policies to supposedly deal with the wave of S&L failures that could be clearly seen on the horizon. According to Lowy, the Regan formulation believed:

1. The current problem is interest rates. High interest rates are due to inflation, which the administration is going to cure. Therefore the problem is temporary.

2. The problem is basically a liquidity problem caused by interest rate regulation. If rates are deregulated, the S&Ls will be able to attract funds. Therefore, rates should be deregulated.

3. There is no money in the budget for bailouts. (Reagan had been against the bailouts of New York City and Chrysler.) Therefore, if S&Ls need assistance, it must be purely paper assistance that has no budgetary cost.

4. The important thing is to pass real deregulatory legislation to give S&Ls the same powers as commercial banks.

The Reagan administration didn't believe that balance sheet insolvency (negative net worth) was significant. As long as an S&L could get enough new deposits to continue in business, the Treasury people believed that it didn't need to be closed. However, in such cases the S&Ls were operating a modified Ponzi scheme because the funds from new deposits were used to pay out account interest; income from mortgages was insufficient. Unwittingly, the Reagan administration propped up Ponzi schemes.

The Reagan administration followed a set of beliefs in almost a religious way; they did not allow reality to interfere with their hypotheses.

In mid-1981, 75% of S&Ls were sure to lose money. About 50 were insolvent and another 300 were sure to become insolvent in the next year. Over 1,000 S&Ls could not meet traditional requirements for net worth as a percentage of assets. Even the healthiest S&Ls would be insolvent in two years if interest rates didn't turn down. For these reasons, it appeared likely that the FSLIC would have to liquidate several hundred billion dollars (about half) of S&L industry assets, with a net cost to taxpayers of about $50 billion to $100 billion. The FSLIC only had assets of $6 billion, and Mr. Reagan was very chary about spending even that amount.

Lowy described a variety of accounting chimeras that were used to try to stem this tide, at least temporarily. He summarized this by saying:

If you now have the impression that bank and S&L regulation was being contorted in an attempt to deal with the S&L problem without spending money, you are correct.... Even with the fancy footwork, there were too many insolvent institutions for the FSLIC to handle with its limited resources.

Therefore, the Reagan administration decided to change the definition of *net worth* so fewer S&Ls would have to be declared insolvent. Lowy described "these shenanigans" as "a strategy to fool the public." One provision (of many) allowed an institution to mark up the value of a property on their books if the property increased in value, but allowed them to keep the original price on their books if the property decreased in value.

The Garn–St. Germain act of 1982 was enthusiastically endorsed by Mr. Reagan. This act provided for extensive deregulation of S&Ls. Amongst other things, it provided the following:

- It essentially swept away almost all state rules and regulations governing S&Ls.[29]

- It allowed S&Ls to convert from mutual banks to stockholders' entities.

- It eliminated the statutory loan-to-value (LTV) tests for making home loans and apartment loans (essentially any down payment was now acceptable, including zero or negative).

- It eliminated the requirement that commercial real estate loans be made on the security of first liens.

- It increased the percentage of assets that an institution could invest in commercial mortgage loans from 20 to 40%.

- It authorized unsecured business loans with up to 10% of an institution's assets.

- It increased consumer lending authority from 20% of assets to 30% of assets.

- A regulation requiring a S&L to have 400 stockholders with no one owning more than 25% of the stock was changed in April 1982 to allow a single shareholder to own a thrift.

- Originators of new S&Ls were allowed to start (capitalize) their S&L with land or other non-cash assets rather than money. (This provision was a boon to land developers who had extra land lying around that they had not been able to develop or sell.)

- S&Ls were permitted to make real estate loans anywhere. They had until now been required to loan on property located in their own market area, with an emphasis on community home building and ownership. But with this new regulation S&Ls were allowed to loan on property too far from home to monitor properly.

During the early years of the Reagan administration, responsibility for the unfolding thrift crisis lay with the Cabinet Council on Economic Affairs, chaired by Treasury Secretary Donald Regan.... Firm believers in *Reaganomics*, this group crafted the policies of deregulation and forbearance and adamantly opposed any governmental cash expenditures to resolve the S&L problem.[30] Furthermore, the administration did not want to alarm the public unduly by closing a large number of S&Ls. Therefore, [they used] FSLIC notes and other forms of forbearance that did not have the immediate effect of increasing the federal deficit. The free-market philosophy of the Reagan administration also called for a reduction in the size of the federal government and less public intervention in the private sector. *As a result, during the first half of the 1980s* the federal banking and thrift agencies were encouraged to reduce examination staff, even though these agencies were funded by the institutions they regulated and not by the taxpayers. This pressure to downsize particularly affected the FHLBB, whose budget and staff size were closely monitored by OMB and subjected to the congressional appropriations process. The free-market philosophy affected not only regulatory and supervisory matters but also thrift and bank chartering decisions.... The devastating consequences of adding many new institutions to the marketplace, expanding the powers of thrifts, decontrolling interest rates, and increasing deposit insurance coverage, coupled with reducing regulatory standards and scrutiny, were not foreseen.[31]

THE FALSE SPRING OF 1983

Lowy wrote a chapter in his book with the title "The False Spring of 1983." He explained the false optimism that prevailed in 1983. Interest rates had come down, and the Reagan tax cuts spurred the economy. Optimism was widespread. A 1983 book on the S&L crisis described it as "resolved." The S&Ls were essentially unregulated. They could invest as they saw fit with funds borrowed below the market due to the deposit insurance guarantee. There was the appearance that all was well. While the S&L industry had lost $6 billion in each of years 1981 and 1982, the industry reported a profit of $2.5 billion in 1983. However, this claim derived from what Lowy described as "funny money accounting."

Despite the reduction in interest rates, most S&Ls still had an inadequate spread between average mortgage interest and average depositors' interest, and were losing money from operations.

The only way out seemed to be to rapidly acquire huge volumes of new loans that would dilute the overhang from past loans, and thereby repair the balance sheet.

The figures for the S&L industry in 1983 provided by Lowy are as follows:

- Bank operations: $4 billion loss.

- Appreciation of assets (from lower interest rates): $2.5 billion profit.

- Fees for newly issued loans: $4 billion profit.

Thus, the reported "profit" was $2.5 billion.

The appreciation of assets was real, although it was a paper gain and added no cash flow. Where the S&L industry made its profit was by issuing a huge volume of new loans in this unregulated atmosphere. The loan fees collected up front were treated as profit; however there was a good deal of "funny money accounting" involved in this. We will discuss this further below.

The real state of the S&L industry was revealed by Lowy:

- Almost 50% of S&Ls lost money in 1983.

- Even with the drop in interest rates the "spread" was inadequate for most S&Ls.

- Use of conservative accounting procedures would have led to a far more pessimistic picture.

The rapid growth of loans led to a good deal of inefficiency, and many of the loans were very risky. The S&Ls were stampeding to acquire loan origination fees, and the safety and security of the loans were often grossly inadequate. Many S&Ls moved out of single-family residence financing to finance complex large developments but they had neither experience nor knowledge to do this effectively.

Lowy emphasized that "construction loans for condominiums, office buildings, shopping centers and hotels require entirely different expertise from single-family lending." S&L managers were typically inexperienced in construction lending and were lured into faulty projects by the high fees paid up front. Lowy also pointed out: "Although the majority of bad construction loans were committed in 1982,

1983 and 1984, the outside auditors and examiners ... usually didn't learn that the loans couldn't be repaid until 1985, 1986 or 1987." With interest rate deregulation, S&Ls had access to unlimited funds, because they could pay whatever was required to attract depositors, who did not worry because the FDIC backed their deposits. When they loaned these funds out, they collected significant loan fees that aided their earnings, under new accounting rules. But unfortunately, many of the loans were bad loans. Lowy described these loan fees as a sort of drug on which the S&Ls binged, while they accrued an increasing supply of poorly conceived construction loans on their books.[32]

In the worst cases, the S&Ls put up the entire cost of the project, including the loan fee, which they paid to themselves and called it income. This created the illusion of profitability, and more funds were attracted via deposits. While these risky loans were made, very little, if any loan loss reserves were put aside. Such processes bordered on the edge of being Ponzi schemes. Lowy's summary was:

> Using the loan fee income, the fast growing S&Ls paid big dividends to stock-holders, paid big salaries and bonuses to management, and built up fleets of airplanes and other luxuries. The excessive lending for office buildings, con-dominiums, and shopping centers led directly to the devaluation of the proper-ties that were being built. Some of the loans involved fraud and some involved dishonest appraisals. But even honest appraisals will be totally wrong if the amount of property built in the marketplace significantly exceeds demand, because without demand, prices will fall precipitously, as they did in Texas, Colorado, and Arizona, wiping out substantially all of the S&Ls in those states. We can't blame all of this on permissive accounting, but if the accounting had been done right, the problem would have been much smaller. Some of the Texas and California high flyers, such as Vernon and Independent American, added a little fraud to the aggressive accounting by having borrowers pay additional fees to *service corporation* subsidiaries, usually in exchange for fictitious *mortgage banking* or other 'services'. This could bring the total fee to 6 or even 10 percent. Of course the S&L financed these fees as well – and what did the developers care, if they were getting a loan of 110 percent or 120 percent of cost without recourse? They already had their profit without having invested a penny.[33]

Loan fees weren't the only way that S&L income was being inflated. Lowy discussed other accounting legerdemain. He also discussed the high leverage inherent in S&L funding of real estate ventures.

THE "GO–GO" PERIOD

The savings and loan industry changed swiftly and dramatically after the deregulation of asset powers and interest rates. The period from 1982 to 1985 was characterized by extremely rapid growth. S&L growth was fueled by an influx of deposits into institutions willing to pay above-market interest rates. In 1983 and 1984, more than $120 billion in net new money flowed into savings and loan associations. With money flowing so plentifully, risk takers gravitated toward the S&L industry, altering ownership characteristics. As the FDIC Report said:

Although more than a few of these new owners engaged in highly publicized cases of fraudulent activity, many others were just greedy.

Sharp entrepreneurs took [advantage of] the large potential profit from owning an S&L, whose charter now allowed a wide range of investment opportunities without the corresponding regulation of commercial banks. Little capital was required to purchase or start an S&L, and the growth potential was great. A variety of non-bankers entered the S&L industry, ranging from dentists, with no experience in owning financial institutions, to real estate developers, who had serious conflicts of interest. To gain entry into the S&L industry, one either acquired control of existing institutions (many of which had converted from mutual to stock) or started de novo institutions. Between 1980 and 1986 nearly 500 new S&L charters were issued, with more than 200 of these issued in just two years –1984 and 1985.... Another major change resulting from deregulation was that, beginning in 1982, S&L investment portfolios rapidly shifted away from traditional home mortgage financing and into new activities. This shift was made possible by the influx of deposits and also by sales of existing mortgage loans. By 1986, only 56 percent of total assets at savings and loan associations were in mortgage loans, compared with 83% in 1978.

In some states, direct investments in real estate, equity securities, service corporations, and operating subsidiaries were allowed with virtually no limitations. S&Ls invested in everything from casinos to fast-food franchises, ski resorts, and windmill farms. Other new investments included junk bonds, arbitrage schemes, and derivative instruments. It is important to note, however, that while windmill farms and other exotic investments made for interesting reading, high-risk development loans and the resultant mortgages on the same properties were most likely the principal cause for thrift failures after 1982. A large percentage of S&L assets were devoted to acquisition, development, and construction (ADC) loans; these were very attractive because of their favorable accounting treatment and the potential for future profit if the projects were successful. The entry of so many S&Ls into commercial real estate lending helped fuel boom-to-bust real estate cycles in several regions of the country.

Interest rates on construction loans are much higher than on other forms of lending; and regulatory accounting practices allowed S&Ls to book loan origination fees as current income, even though these amounts were actually included in the loan to the borrower. For example, a developer might have requested a $1 million loan for two years for a housing development; the institution might have charged four points for the original loan and 12 percent annual interest. However, instead of requiring the borrower to pay the interest for two years ($240,000) and the fee ($40,000), the S&L would have included these two items in the original amount of the loan (which would have increased to $1.28 million). . . . There are many notorious examples of how this system was abused by unscrupulous S&L owners reporting high current income on ADC loans while milking the institution of cash in the form of dividends, high salaries, and other benefits. A rapidly growing S&L could hide impending defaults and losses by booking new ADC loans.[34] The rush into construction lending by S&Ls was such that among the fastest growers, loan fees accounted for substantially all net income in the crucial years 1983 and 1984. Moreover, although the majority of S&Ls were not fraud-ridden, few had the management expertise necessary for dealing with the new lending opportunities, particularly the inherently risky ADC lending. In many cases, prudent underwriting standards were not observed, and the necessary documents and controls were not put in place. Lending on construction projects was appraisal driven and was often based on the overly optimistic assumption that property values would continue to rise. S&Ls sometimes loaned the entire amount up front, including interest, fees, and even payments to developers, but did not check to ensure that projects were being completed as planned. Moreover, S&L ADC loans frequently were non-recourse: the borrower was not required to sign a legally binding personal guarantee.[35]

Another factor in the S&L problem was that more and more, mortgages were packaged into large investment vehicles that isolated homeowners from the institutions holding their mortgages. (See for more on "Mortgage-Backed Securities" in Chap. 2). Lewis[36] pointed out that Congress passed a significant tax break for S&Ls in 1981 that allowed them to sell off bad loans at a discount but amortize that loss over the life of the loans, thus putting a rosy view on their balance sheet that was not justified. The S&Ls were required to invest the proceeds from such "fire sales" in new loans at a higher rate. Thus, S&Ls bought bad loans from one another at discounts, amortized their losses, and reported better balance sheets than they deserved. The sale and purchase of huge numbers of mortgages created a bonanza for investment bankers who packaged these mortgages into investment vehicles, thus establishing the initial foundation for the sub-prime boom that was to follow twenty years later.

FRAUD AND MISCONDUCT

Pizzo, Fricker and Muolo (PFM) [37] described the "looting of America's S&Ls" in 500 pages. We will not attempt to review this enormous record of bad management, deceit, and criminality, but we will merely be content to briefly mention a few examples. The reader is referred to PFM for many more details.

PFM described the evolution of tiny conservative Centennial Savings and Loan in a small town (Guerneville, population 1,700) in Northern California that began with an investment of $2 million in 1977 for the purpose of supplying home loans in the hope that Guerneville would start growing like its bigger neighbor, Santa Rosa. However, growth was slow. With the impending prospect of deregulation in sight, Centennial hired a "go-go" man as its president in 1980. He was able to acquire large amounts of money from brokered deposits at above-market interest rates, and proceeded to launch Centennial into the construction business by paying an exorbitant sum for a local construction company, and went on from there. According to PFM, Centennial purchased a stretch limousine in the president's name, bought 25 luxury cars for management use, leased a twin-engine turboprop airplane, bought and remodeled an office building for itself at a cost of $7 million, bought a property from the president for 10 times what he paid for it, hired a European chef full-time, and paid for lavish trips, furniture and remodeling of property owned by the president. A Christmas party in 1983 cost $148,000. The president and the chairman of the board declared $800,000 bonuses for themselves in 1983 – totaling 2/3 of the reported income of Centennial. Meanwhile, Centennial was dealing in land and property. Centennial hired retired regulators with strong connections "to calm the regulators down." Everything was working fine for a couple of years. Self-dealing was rampant.

But in 1985, regulators finally closed down Centennial as insolvent. At that time, Centennial had swelled to $404 million in assets (compared to about $2 million three years prior). Eighty percent of the $435 million in deposits were high-cost brokered CDs (it is normal that this should be less than 20%). Thirty-six percent ($140 million) of Centennial's loans were tied up in high-risk ventures owned by cronies or subsidiaries of Centennial. It took a couple of years for the federal investigators to sift through the rubble. Finally, 26 charges were filed by the FBI against the former president of Centennial, but he died not long after — some say by suicide. The final bill to the public was $165 million.

Although there were many other S&Ls that lost more money, the spectacular rags-to-riches rise of Centennial in just three years is remarkable.

Perhaps the biggest S&L failure was Lincoln, the acquisition of Charles Keating's American Continental Corporation (ACC). This is a long, involved story that is difficult to summarize.

Charles Keating was a Cincinnati attorney who began his career working for a wheel-a-deal financier named Carl Lindner, who invested in a variety of subsidiaries in his domain. His Phoenix-based home building subsidiary was having problems, and Keating bought it from Lindner in 1978 and renamed it *American Continental Corporation* (ACC). Keating was a lifelong foe of pornography and it is often quipped that his ethics were very different in the bedroom and the boardroom. He and Lindner were accused of misusing S&L's funds between 1972 and 1976, and he accepted the 1979 judgment without admitting guilt. In the early 1980s Keating was managing a home building business in Phoenix. But Keating had much greater ambitions.

He acquired a good deal of cash ($100 million) by selling junk bonds and stock for his ACC through Drexel as underwriter. With this, he paid $51 million in cash for Lincoln Savings and Loan based in Irvine, California. In acquiring Lincoln, Keating "assured regulators in writing that 'no changes are expected in the performance of the institution' regarding home lending." However, as soon as he secured control of Lincoln, he "began using Lincoln's money to invest in stocks, bonds, and high-risk loans on speculative ventures run by ACC subsidiaries." Eventually Lincoln invested about $800 million (10 percent of its portfolio) in junk bonds – mostly bought through Drexel. PFM described the shenanigans at Lincoln:

> Keating replaced Lincoln management with American Continental Corp. employees. He made his son (28 years old and without a college degree), chairman of the board and head of Amcor, ACC's key development subsidiary. Charles III was paid about $800,000 a year. Later the younger Keating, testifying at a congressional hearing, admitted that he would sign his name to anything submitted by his father and two other top executives at ACC. Besides hiring a covey of relatives, Keating also liked to pay large salaries to loyal secretaries, some earning as much as $100,000 a year. Keating also paid himself well. Regulators later said that during the five years that he ran Lincoln – which represented 90 percent of his holdings, he and his family collected $41.5 million in salary, benefits, and perks.[38]

Keating's connections included a number of politicians as well as some shady characters (these are not necessarily exclusive). He provided more than $97 million to John Connally and his partner, former Texas Lieutenant Governor Ben Barnes.[39] But Keating had another side to his personality. He provided millions of dollars to his favorite charities.

Keating's life style became flamboyant. According to PFM:

At Keating's swank American Continental offices on Camelback Road in Phoenix, banks of computers monitored financial markets worldwide; the company had private jets and a helicopter; Keating had two vacation homes on Cat Cay in the Bahamas; he spent more than $1 million on professional football tickets; and over a five-year period, he and his family and friends used Lincoln Savings' money to spend 263 days traveling by private aircraft in Europe.

Keating began work on the $265 million edifice that would become the symbol of his empire, the 130-acre Phoenician Resort at the base of Camelback Mountain in Phoenix. Keating personally oversaw the gold-and-marble construction to make sure his guests – who would have to pay up to $500 a night – would get their money's worth. . . . The hotel's amenities were legion: a 100-foot water slide, 18 grand pianos including nine Steinways, a pool lined with mother-of-pearl, a 32-person hot tub, acres of pools and golf courses, a nightclub called Charlie-Charlie's, numerous restaurants including one entered through a waterfall, 125 South Sea islanders from Tonga imported as grounds keepers, 1,500 full-grown palm trees trucked in from Florida, and gilded ceilings hand-painted by 'an old friend from Europe', according to Keating.

It is instructive to note the excerpts from the diary of Doug Doolittle, a young (and inexperienced) Special Projects Manager of Lincoln over a critical period in Lincoln's history. Only a few short passages are reproduced here: [40]

July 23, 1986 – Life is great! It is 1986, Ronald Reagan is coming to the end of one of the most successful presidencies in recent memory, the American economy is booming, the stock market is continually reaching all-time highs.

When ACC acquired Lincoln, Lincoln's main business was in residential mortgage loans. These were safe investments but boring! Mr. Keating changed all that and by using his expertise in real estate development and his contacts from ACC, has shifted Lincoln's main activity to land development projects. He isn't even risking Lincoln's depositors' money since all deposits up to $100,000 are insured by the government. In the last two years, the real estate transactions have provided the main source of Lincoln's profits.

March 26, 1987 – I suggested we try to do a deal for one of the Hidden Valley parcels. I've just received an independent appraisal for one parcel of 1,000 acres in Hidden Valley that values it at $8.5 million. Given that the original cost to Lincoln was $2.9 million, we should be able to realize a substantial profit.

March 31, 1987 – The Hidden Valley transaction went through today. Wescon bought the 1,000 acre Hidden Valley parcel for $14 million! That is an $11.1 million profit!

April 15, 1987 – It seems that Wescon was given an unsecured $3.5 million loan by ECG Holdings at the end of March, just before the Hidden Valley/ Wescon transaction went through. The $3.5 million is the down payment Wescon used for the transaction. The remaining purchase price was paid by Wescon, issuing a note to Lincoln of $10.5 million. Wescon will only pay 10% annual interest on the note (Lincoln's brokered CDs are offering 11%) and only annual payments based on a 20-year payment (with a balloon payment due in six years) schedule need to be made, so Wescon really has been given an amazingly good deal. The note is very unusual – I just hope Wescon can make the payments on it.

April 22, 1987 – I'm beginning to be concerned about the Hidden Valley/ Wescon transaction. I did some checking on Wescon and its net worth is less than $50, 000. How is it going to meet its payments on a note of $10.5 million? If Wescon can't meet these payments, I'm not sure we should have taken credit for the profit on the sale of the 1,000 acres in Hidden Valley. If the Wescon notes aren't worth $10.5 million, then we didn't receive $14 million for the parcel, and then surely the profit on the transaction must be less than $11.1 million.

June 2, 1987 – Discovered an interesting thing today. Lincoln made a loan commitment of $30 million to ECG Holdings at the end of March and $19.6 million was immediately withdrawn in cash.

April 25, 1988 – There were eight transactions involving parcels from Hidden Valley (including the Wescon deal); in total, they contributed $103 million to revenue and $62 million to pretax profit. These deals were all structured like the Wescon transaction with a down payment of 25% from the buyer and notes receivable for the balance. It really helped swing the deals that Lincoln had made substantial loans for other purposes to each buyer – to a total tune of over $200 million.... My only worry is that I've just noticed that the prospectus shows that the market value of ACC's investments is way below their book value.

May 25, 1989 – I haven't written in this journal for over a year and, looking back on what I wrote then, I can't believe how naive I was. I really thought

Keating was a God and could do no wrong! Lincoln finally was seized by the federal regulators last month although the writing was on the wall long before that. Heck, I even had suspicions two years ago, but because I was so dazzled by Keating, I convinced myself that there wasn't a problem. . . . How could I have been so gullible and stupid?

The bank examiners were not unaware of some of Keating machinations and excess expenditures, but they were slow to react. Chairman Ed Gray of the Federal Home Loan Bank Board struggled to understand how big a problem he had on his hands.

His examiners in the field were telling him that the situation was bad and getting worse – while industry "experts" were saying that the problems were temporary, and were nothing to worry about. Gray insisted that limits needed to be put on the proportion of direct investments (as opposed to mortgage loans) that S&Ls could make with federally insured deposits. In December 1984 he had proposed limiting direct investments to 10% of assets, and the Federal Home Loan Bank Board approved the new regulation in January 1985. It was set to go into effect in mid-March (retroactive to December 10, 1984). Rogue savings and loan entrepreneurs were furious, especially at Lincoln Savings and Loan where the main reason for having an S&L was to gain access to its deposits to fuel speculative investments. PFM described Keating's investments in his first year at Lincoln that included (amongst others):

- $18 million in a Saudi bank

- $2.7 million in an oil company

- $5 million in junk bonds

- $132 million in a takeover bid

- $19.5 million in a hotel.

Keating "completely abandoned the home loan market, turning instead to invest-ment speculation." (In 1985 Lincoln originated only 11 mortgages, and 4 were for employees.) Keating would later tell a judge: "home loans were not his thing."

When Ed Gray wanted to limit direct investment rights, Keating began a bitter personal vendetta against Gray.[41]

Keating hired economist Alan Greenspan as a consultant to Lincoln Savings to fight the battle for more direct investments. PFM described Greenspan's role. Greenspan wrote to Gray "that deregulation was working just as planned." Greenspan named 17 S&Ls including Lincoln that had reported record profits and were supposedly prospering under deregulation. Four years later, 16 of the 17 S&Ls Greenspan had mentioned in his letter were defunct.

> "In February 1985 Greenspan again wrote the Federal Home Loan Bank in San Francisco on behalf of Keating, arguing that Lincoln should be exempted from the direct investment limit. A week later he testified before a House subcommittee that direct investments were sound investments for S&Ls."[42]

Despite efforts by Keating, Greenspan, and others, direct investment limits went into effect in March 1985, and many S&Ls did not meet the requirements, including Lincoln.

In 1987 Greenspan was appointed chairman of the Federal Reserve Board, perhaps as a reward for his astute observations regarding direct investments by S&Ls. Continuing his brilliant insights in a long career as head of the Fed, he single-handedly used monetary policy to propel stock market and real estate bubbles that eventually popped. Rarely has such respect been accorded to such an incompetent.

The S&Ls had plenty of money (the depositors' that is) and they were not unwilling to distribute it to politicians. It is therefore not surprising that a number of members of Congress opposed constraints on direct investments.

But the real problem was that Gray did not have adequate staff to investigate and enforce the regulations, and what staff he had were grossly underpaid, and in many cases lacking competence. He went to the Reagan administration with hat in hand to ask for a doubling of his staff and a significant increase in salary. The response according to PFM was as if it were taken right out of *Oliver Twist*: "You want **MORE** examiners?"

Gray was insistent and persistent in his intent to increase the number and pay of examiners. But he was under constant attack by members of Congress, who likely were paid off by S&Ls, or at least had investments in S&Ls. Keating tried to buy Gray off by offering him a job as "president without duties."

In March 1987, after several years of freewheeling spending and bad economic choices, Lincoln Savings and Loan was in deep trouble and about to go under. In order to protect the investors in Lincoln, federal regulators were considering taking over the company. Keating would have none of this, of course, and thus he decided to collect on his investments in Washington. In late March 1987, Keating set up a meeting with one of his closest associates in Washington, Senator Dennis DeConcini, the Democratic senior senator from Arizona. Keating requested that DeConcini set up a meeting with the federal regulators, with the purpose of getting them to leave Lincoln Savings and Loan alone. DeConcini was quite willing to follow up on the request, since Keating had donated thousands of dollars to DeConcini's senate campaigns. So DeConcini sought out a number of senators that Keating had donated money to in the past and invited them to a meeting with the regulators on April 2.

Over the next several months, Lincoln Savings and Loan continued its death spiral, eventually falling apart in early 1989. When the final tallies were counted, roughly $3.2 billion was lost by the corporation, including $2 billion in investor money; the investments were bailed out by the government through the FSLIC.

Without going through all the sordid details, two trials were held, one in 1991 and one in 1993. In the first trial Keating was found guilty on 17 of 18 counts of securities fraud; in the second trial he was found guilty on 73 counts. He spent 50 months in prison and was fined several hundred thousand dollars (which seems to be a drop in his bucket).

Charlie Keating always took care of his friends, especially those in politics. McCain was no exception. In 1982, during McCain's first run for the House, Keating held a fund-raiser for him, collecting more than $11,000 from 40 employees of American Continental Corp. McCain would spend more than $550,000 to win the primary and the general election. In 1983, as McCain contemplated his House re-election, Keating hosted a $1,000-a-plate dinner for him, even though McCain had no serious competition. When McCain pushed for the Senate in 1986, Keating was there with more than $50,000. By 1987, McCain had received about $112,000 in political contributions from Keating and his associates. While in the House, McCain, along with a majority of representatives, co-sponsored a resolution to delay new regulations designed to curb risky investments by thrifts such as Lincoln. In the end, McCain received only a mild rebuke from the Ethics Committee for exercising "poor judgment" for intervening with the federal regulators on behalf of Keating. Still, he felt tarred by the affair – and well he should. But that did not stop him from running for president in 2008.[43]

As Lowy pointed out:

Reports of fraud and misconduct by S&L owners and officers, more than the amount of money that taxpayers would pay, is what has made the

American public angry about the FSLIC bailout. The large sums of money are not comprehensible. But the idea that the S&L managers who portrayed themselves as pillars of society made off with the loot excites people.

Lowy distinguished between different types and degrees of misconduct. He claimed that the amount of fraud and misconduct was exaggerated and "lumped together crimes of varying degrees, regulatory violations, mismanagement, and personal aggrandizement by S&L officers." Nevertheless, over 2,000 criminal referrals were made in S&L failure cases.

Nine categories of "insider misconduct" were identified including payment of exorbitant personal expenses (not necessarily illegal), preferential loans to companies affiliated to insiders (also not necessarily illegal), inaccurate financial reports, and acceptance of false information submitted by borrowers. Lowy pointed out that while this behavior is reprehensible and should be prosecuted, "what kills banks – and what made S&Ls so deeply insolvent – is bad loans." Bad loans were sometimes made by bad managerial decisions. However, in some cases, bad loans were the end-product of either (1) payoffs to insiders, (2) loans to associates of insiders, or (3) loans to fraudulent borrowers whose misrepresentations were not adequately checked by the S&Ls.

Type 1 loans are clearly and obviously illegal, and often were responsible for the greatest S&L losses. Lowy claimed that investigators of the S&L scandal:

> ... tended to lump together Types 1, 2, and 3 fraud and misconduct with a host of other types of insider misconduct, such as keeping inaccurate records, filing incorrect reports, paying themselves too much, and spending too much corporate money entertaining themselves and their customers. These other types of insider misconduct, while they do evidence the kinds of attitudes that lead to laxness and mismanagement – and often accompany Type 1 and Type 2 misconduct – are even more frequently symptoms that there is no real capital at risk for the owners and directors to protect. Expense account liberties and high corporate living, while repugnant to low-paid regulators – and especially repugnant after an institution's failure – are almost never of a sufficient magnitude to cause insolvency. David Paul's excesses at CenTrust, including buying an art collection, a yacht, a sailboat, Limoges china, and Baccarat crystal, didn't lose more than $20 million (after selling them all at auction) against an insolvency of $2 billion – less than 1 percent. It sounds pretty sensational, but it is not the heart of the problem.

Lowy claimed that another type of misconduct "probably cost more than all the others put together, even though it didn't cause any failures." This involved concealing loan defaults with phony transactions and filing false reports.

Finally, Lowy asked:

> Why was there so much fraud at Texas S&Ls? Substantially all of the banks there failed, too, yet there have been few allegations of fraud by bankers. Were the bankers better people? Or were there other reasons why fraud was prevalent in the S&Ls and not in the banks?

Lowy's answer to this question seems to be too muted. The real answer seems to be because they were not prevented from doing it.

THE AFTERMATH

It is now clear in retrospect that a large portion of the funds eventually paid out by the federal government to bail out failed S&Ls could have been avoided by timely and effective action before the problem escalated out of hand. The Reagan administration simply could not face reality and Congress was no better.[44] According to Lowy:

> Practically no one in Congress could conceive of tax dollars being used to pay for this problem. They got angry if you suggested this possibility. . . . Speaker Wright and powerful members of the Banking Committee were saying $5 billion or so was plenty.

In 1987, after auditors said that at least $50 billion was needed, some Congressmen reluctantly proposed to raise the $5 billion budget to $15 billion. It was voted down 258–153.

Lowy in reviewing the policy of "keeping zombie S&Ls open" put it very succinctly: "the delay cost a whole lot."

The FDIC Report suggested that it is:

> . . . amazing that such a monumental crisis, and one given top priority by the new administration, had been virtually ignored as an issue during the 1988 presidential campaign. This invisibility . . . was partly due to the continued reluctance to admit that taxpayer dollars would be required, and

partly to the fact that members of both political parties were vulnerable to criticism for their role in the crisis.

The FDIC Report concluded:

> It must be concluded that the savings and loan crisis reflected a massive public policy failure. The final cost of resolving failed S&Ls is estimated at just over $160 billion, including $132 billion from federal taxpayers and much of this cost could have been avoided if the government had had the political will to recognize its obligation to depositors in the early 1980s, rather than viewing the situation as an industry bailout. Believing that the marketplace would provide its own discipline, the government used rapid deregulation and forbearance instead of taking steps to protect depositors. The government guarantee of insured deposits nonetheless exposed US taxpayers to the risk of loss – while the profits made possible by deregulation and forbearance would accrue to the owners and managers of the savings and loans.

> The person most responsible for the depth and extent of the Savings and Loan Scandal of the 1980s was President Ronald Reagan.

His policy, implemented by Treasury Secretary Regan, was to essentially interpret "deregulation" of banks as no regulation. As a result, the banks were allowed (perhaps even encouraged) to run wild with investment schemes backed by the FSLIC that would have been considered unimaginable in prior years. Reagan's legacy included the cost to taxpayers of over a hundred of billion dollars to bail out the banks, and a large escalation in the federal debt. Reagan's theories of trickle-down wealth and increased revenues from lowered taxes were phony from the beginning. Reagan was a terrible president. Yet, Reagan is revered by Republicans, and in the 2008 Republican primary contest, Republican candidates vied with one another for the right to claim to be the most like Reagan!

The FDIC Report derived the following regulatory lessons of the S&L disaster:

- First and foremost is the need for strong and effective supervision of insured depository institutions, particularly if they are given new or expanded powers or are experiencing rapid growth.

- Second, this can be accomplished only if the industry does not have too much influence over its regulators and if the regulators have the ability to hire, train, and retain qualified staff. In this regard, the bank regulatory agencies need to remain politically independent.

- Third, the regulators need adequate financial resources. Although the Federal Home Loan Bank System was too close to the industry it regulated during the early years of the crisis and its policies greatly contributed to the problem, the Bank Board had been given far too few resources to supervise effectively an industry that was allowed vast new powers.

- Fourth, the S&L crisis highlights the importance of promptly closing insolvent, insured financial institutions in order to minimize potential losses to the deposit insurance fund and to ensure a more efficient financial marketplace.

- Finally, resolution of failing financial institutions requires that the deposit insurance fund be strongly capitalized with real reserves, not just federal guarantees.

However, these do not seem to go far enough. What needs to be added is:

As long as the government agrees to bail out the deposits of failed banking institutions, the government has the right and responsibility to supervise these banks, to assure that they follow conservative investment and accounting practices, and that improper or fraudulent policies are quickly recognized and dealt with. Considering the sub-prime mortgage fiasco of 2007–2008, it is clear that neither the Government nor the public has yet learned the lesson from the 1980s; deregulation is still rampant, and still interpreted as no regulation.

THE BULL MARKET OF 1982–1995

Maggie Mahar described the bull market of 1982–1999 in a lengthy volume.[45] The years 1966 to 1982 represented 16 years of poor performance by the stock market. During much of this period, inflation outpaced the return from investment in stocks, and one did better by investing in bank CDs that were paying very high rates with the safety feature of backup from the FSLIC and the FDIC. In 1980–1982, the Dow-Jones average was trading at around seven times earnings. But as Mahar said: "one of the peculiarities of Wall Street is that buyers shun a bargain." Or put differently, buyers rely more on momentum than on value.

This had been going on for so long that by the early 1980s, younger people had never experienced a real bull market.

Ronald Reagan was elected in 1980. His impact on the economy was significant. Reagan had a belief system that he followed religiously, and ignored the "conventional wisdom" as well as reality. Reagan believed (or at least claimed that he did) that by decreasing taxes, particularly for the rich, government revenues would grow because business activity would overwhelm the reduction in tax rate. He also believed that this expanded wealth at the top would "trickle down" to the lower echelons via expanded employment opportunities. In addition, Reagan was vehemently opposed to almost any form of government regulation of anything, particularly banks, and seemed to desire to reduce the US government to a minimal role in everything except the military, for which he had no limits. Under Reagan, the maximum income tax in the highest bracket was reduced from 70 to 28%, and the capital gains tax rate was reduced from 30% to less than 20%.[46] And Mr. Reagan flooded the money supply with an endless supply of cash. Although his policies did not work the way they were supposed no (the federal deficit grew at an unprecedented rate (see Figure 2.5)), his policies did contribute to the beginning of a gigantic boom in the stock market since interest-bearing investments became less attractive as interest rates came down. Although the "conventional wisdom" would have suggested that low taxes, a large federal deficit, and an expanded money supply should have produced rampant inflation, Mr. Reagan had never taken Economics 101 and therefore was not aware that according to basic theory, his approach should not work. Like the nearsighted Magoo crossing a busy street with traffic zooming past him, but never hitting him, Mr. Reagan defied the laws of economics and won – at least to the extent that inflation was not rampant, the economy recovered, and the stock market boomed. We did have the S&L crisis, and deficits soared, but those costs were paid for after Mr. Reagan left office.

Martin Lowy put it very succinctly.[47] In late 1982, the Fed eased its monetary policy and interest rates began to decline. The stock market reacted by initiating the greatest bull market in history. Real estate also responded with a boom of its own, as Lowy described:

> ... propelled in part by tax legislation passed in 1981 that permitted investors in real estate to take accelerated depreciation and to deduct interest paid on their personal tax returns. ... Real estate ventures multiplied. Pent-up demand after years of high interest rates made fortunes for single-family-home developers. The great god mammon was on the loose, as greed became socially acceptable, investment banking became the most popular professional aspiration for college

seniors, and everyone thought that he or she not only could, but would and should be rich. If there ever had been strong morals in American business – which had a spotty record at best – they gave way to the urge to take advantage of the moment.

The stock market took off in August 1982, and began a spectacular 17-year rise to 1999, with a few hiccups along the way (including the crash of 1987 – see "The Crash of 1987" in Chap. 2). The S&P 500 Index is shown in Figure 2.3. Even more spectacular is the dot.com era within the 1982–1999 bull market, as illustrated in Figures 2.4, 2.5 and 2.6.

Many treatises have been written to explain this gigantic long-lived bull market. Some say that it was at least partially an over-reaction to the under-pricing of stocks in the 1970s. Others say that except for some wild dot.com stocks, the overall stock market was not actually overpriced in the 1980s and 1990s.[48] Some explain it in terms of falling interest rates that made savings and investments in interest-bearing securities far less attractive. A plentiful money supply and enthusiasm from the Federal Reserve chairman did not hurt.[49] Income taxes and capital gains taxes were sharply reduced during this era, while government deficits soared. Almost all of the stock market analysts were wildly optimistic.

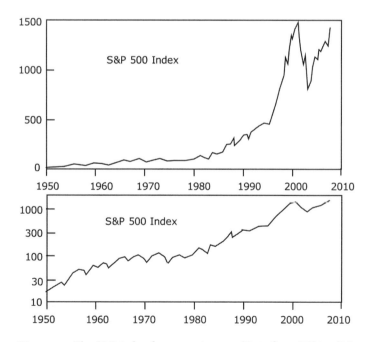

Figure 2.3 The S&P index from 1950 to 2007 (Data from Wikipedia).

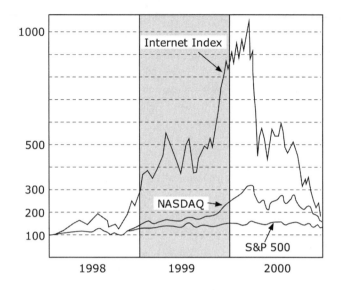

Figure 2.4 Market indices during the final phase of the dot.com boom/bust (based on setting indices to 100 at the start of 1998). (Adapted from "Dot.Con", John Cassidy).

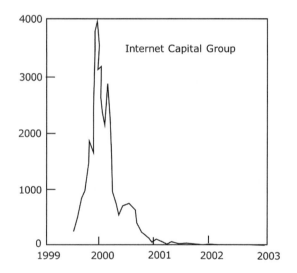

Figure 2.5 Stock price history of Internet Capital Group. Today, you can buy a facsimile stock certificate (suitable for framing) for $69.95. (Adapted from "Dot.Con", John Cassidy).

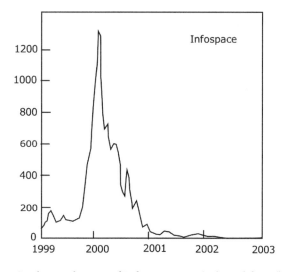

Figure 2.6 Stock price history of Infospace.com. (Adapted from "Dot.Con", John Cassidy).

The advent of 401(k)-retirement plans poured billions into the stock market. There was a widespread and pervasive belief that stocks were proven to be the best long-term investment,[50] and there was a concomitant belief that there would quickly be a recovery from any temporary downturn. Mergers and acquisitions, and stock buyback programs reduced the number of shares available.[51] Big gains were possible from corporate takeovers,[52] and the advent of junk bonds for this purpose expanded the opportunities. At about the time that the stock market had built up a strong upward momentum, the advent of the dot.com revolution poured fuel on a raging fire, driving the flames to unprecedented heights. Initial public offerings were greeted with wild enthusiasm. Viewpoints on how and why to value stocks became much more subjective and all the old standards were discarded. All of the above factors undoubtedly played a role in fostering the great bull market of 1982–1999. But the key factors in continuation of the bull for 17 years seem likely to be more psychological than logical, fiscal or fundamental. The large numbers of "baby boomers" that were approaching middle age without much security for the future demanded wealth, and they weren't going to get it from their salaries. Paper profits on assets were their only hope and they pinned their futures on that hope. This generation has been characterized as the "something-for-nothing" genera-tion. They had no sense of "value" and were willing to bid any piece of paper up to astronomical heights. Once that momentum was established, there was no force capable of stopping it, and the one institution that might have put on the brakes, the Federal Reserve, ran scared before the possibility of raining on their picnic.

Maggie Mahar emphasized the importance of the huge expansion in 401(k) plans as a source of funds to drive the stock markets upward, starting in the 1980s, and even more so in the 1990s. As Smith explained: [53]

> During the 1990s global pension fund assets grew an average of 15 percent per year, from $4.6 trillion to $15.9 trillion. Equity holdings of those funds jumped from $1.6 trillion to $8 trillion, or from 35 percent 51 percent of total assets. By the end of the decade, the stock holdings of retirement funds made up nearly one-quarter of total global equity market capitalization.

With money flowing freely and taxes down, the advent of very large-scale 401(k) accounts and a widespread belief that stocks were the best long-term investment, the stock market took off in 1982 and reached its ultimate peak in early 2000 with the culmination of the dot.com bubble.

The bull market of the 1980s morphed into the dot.com mania of the 1990s. The dot.com mania is discussed in "The Dot.Com Mania" in this chapter. The end of the dot.com bubble occurred in 2000 as the NASDAQ index dropped by about 75% and many dot.com stocks went out of business. This steep decline signaled a termination of the great bull market of 1982–1999. This is described in "Bursting of the Bubble" in this chapter. However, the Federal Reserve struggled valiantly to reinflate the bubble by dropping interest rates and pouring money into the banks. To some extent, they were successful because the stock markets recovered some of their losses from 2002 to 2007. But the major effect of Federal Reserve actions was to inflate a new bubble in housing and mortgages. This is described in "The Sub-Prime Real Estate Boom 2001–2007" in this chapter.

As is usual with booms and bubbles, there were many learned professors and advisors who provided explanations why the incredible rise in stock prices was appropriate and justified by economics.[54] The media produced a glut of articles arguing that there was a "New Economy - one in which the old rules of economics no longer applied."[55] Cassidy[56] quoted Michael Mandel, a Harvard Ph.D. in economics who served as Business Week's economics editor, who wrote articles entitled "The Triumph of the New Economy, The New Business Cycle, and The New Growth Formula" in which he argued that high-technology was now the driving force in the US economy, "leading to good times for the foreseeable future."

During the dot.com boom phase of the great bull market, it was widely believed that the advent of microelectronics, the personal computer, e-mail, the Internet, the laser, and other advanced technologies would produce what Greenspan said was "a once or twice in a century phenomenon" that would elevate productivity to new levels. To the consternation of market enthusiasts, initial Government estimates of productivity failed to verify the intuitions of market analysts. According to Cassidy, "productivity data by the BLS and Commerce Department in 1996 showed little improvement." Abby Cohen (a leading figure in Wall Street advocating dot.com stocks) said: "I believe that the Government's productivity figures are wrong." As Cassidy said: "She claimed that voice mail, word processing, etc. improved productivity by leaps and bounds." Greenspan also had the same belief. According to Cassidy, "Greenspan believed that American firms and workers were becoming a lot more productive, even if the official statistics were failing to pick this up." In September 1996, "Greenspan called some Fed economists together and asked them to reexamine the productivity figures." Cassidy summarized:

> The staff economists confirmed what a number of academic studies had already found: the reason that the overall figures were so low was that the service sector, which employs about 2/3 of the workforce, had seen virtually no productivity growth at all in three decades.[57]

At a meeting convened by Greenspan, Robert Shiller ("Irrational Exuberance") warned of the bubble in the overvalued stock market. Greenspan was not convinced.

> Since the early 1980s, American firms had been spending heavily on computers and other forms of information technology, investments that should have led to higher productivity growth throughout the economy. The failure to find such a link was known to economists as the 'productivity paradox.' [It was claimed] that the United States was now finally receiving the payback for the investments it had made in computers, [and the] productivity paradox seems to be over.[58]

As it turned out, by about 2000, it appeared that the modern microelectronics age had indeed increased productivity. It is claimed that between 1995 and 1999, labor productivity had grown at an annual rate, of 2.8 percent, double the 1.4 percent rate recorded between 1973 and 1995. Although productivity data are not necessarily reliable, such an increase seems possible.[59] The real question is whether this was a temporary transition to a new plateau, or whether continued increases in productivity would propagate through the 2000s and 2010s. Greenspan continued to take the optimistic view with phrases such as "awesome changes" were taking place in "the ways goods and services are produced and, especially, in the way they are

167

distributed to final users." It does appear that there have been significant advances in productivity, but stock prices have outrun these advances by a wide margin. The increase in paperwork has had a counterbalancing effect on the increase in efficiency. For example, the typical medical office has as many aides and paper shufflers, as it does medical professionals. Government policies require a huge amount of paperwork from all business establishments. All of that work is considered to be part of the Gross National Product. If one person produces a product, and another has to supervise him and file reports, and a third has to keep track of his time and fill out accounting forms, one must wonder whether the reports and accounts are included in "productivity."

Cassidy commented at length on the role of Mr. Greenspan and the Fed in the dot.com bubble (See "Greenspan and the Federal Reserve" in this chapter). According to Cassidy, previous Fed chairmen would have been alarmed at the effervescent stock price increases in 1997:

... but Greenspan didn't seem concerned. Since his 'irrational exuberance' speech the previous December, he had hardly mentioned the stock market publicly. Privately, he still regarded the question of whether there was a speculative bubble as an open one. After making the famous speech, Greenspan had ordered the Fed's staff economists to determine whether there was any objective way to tell when arising market had turned into a speculative bubble. After an extensive survey of past speculative episodes, the best economic brains at the Fed concluded that there wasn't any reliable method. Speculative bubbles could only be identified definitively in retrospect.

In "The Rationality of Experts" in Chap. 1 we discussed the analysis of G. J. Santoni a senior economist at the Federal Reserve, who could not figure out any way to determine whether a bubble is occurring or not. Apparently, repeated doubling of asset prices does not necessarily qualify as a bubble to the Fed.

As Cassidy pointed out, Greenspan was fearful that any weak attempt to rein in the bubble could just as easily produce a runaway crash in stock prices, for which he would be held responsible. Cassidy also argued that low interest rates weren't Greenspan's only contribution to the stock market boom. His frequent references to the benefits of new technology, and his refusal to criticize excessive speculation, also played an important role. In August 1999, Greenspan said stock prices reflected:

...judgments of millions of investors, many of whom are highly knowledgeable about the prospects for the specific companies that make up our broad stock price indexes.

There were a handful of doubters during the dot.com boom. Maggie Mahar described the commentaries of David Tice who started a short-position fund in 1995, Richard Russell's Dow Theory Letter, Marc Faber's Gloom, Doom and Boom report, and many reports by Shiller and Campbell leading up to Shiller's book: "Irrational Exuberance."

THE CRASH OF 1987

Mark Carlson of the Federal Reserve described the market crash:[60]

> On October 19, 1987, the stock market, along with the associated futures and options markets, crashed, with the S&P 500 stock market index falling about 20 percent. The market crash of 1987 is a significant event not just because of the swiftness and severity of the market decline, but also because it showed the weaknesses of the trading systems themselves and how they could be strained and come close to breaking in extreme conditions. The problems in the trading systems interacted with the price declines to make the crisis worse. One notable problem was the difficulty gathering information in the rapidly changing and chaotic environment. The systems in place simply were not capable of processing so many transactions at once. Uncertainty about information likely contributed to a pull back by investors from the market. Another factor was the record margin calls that accompanied the large price changes. . . . Finally, some have argued that *program trades*, which led to notable volumes of large securities sales contributed to overwhelming the system.

Carlson described events that led up to the crash. In the gigantic bull market from 1982 to 1987, the S&P 500 Index rose from about 120 in 1982 to a peak of over 300 in the late summer of 1987. This increase was roughly the same magnitude as the stock market rise from 1924 to 1929 (Figure 1.3 in "Rationality of Experts?" in Chap. 1 illustrates this relationship very clearly). As in the 1920s, takeovers (assisted by favorable tax laws) played an important role in this bull market. The expansion of pension funds also contributed. As this great expansion of the stock market took place, professional investors who managed very large funds developed sophisticated tools to manipulate their investments. These involved *programmed trading* (widely known as *portfolio insurance*) and *index arbitraging*. These are discussed in "Internal Feedback" in Chap. 1. In programmed trading, the goal was to capitalize on upward movements of stock indices by buying into them, and selling into descending markets to avoid greater losses. As "Internal Feedback" in Chap. 1

169

shows, this created positive feedback that amplified market movements. Since many of the major investors used similar algorithms, large market investors were moving in "lock-step." In addition, arbitragers were active in buying and selling when differences developed between current market averages and futures.

In the days leading up to October 19, 1987, the stock market was already weakening substantially. Carlson suggested that this was due to expectations that: (a) tax benefits for takeovers were likely to be eliminated, and (b) increases in the US trade deficit might lead the Fed to raise interest rates. Many analyses of the 1987 market crash abound on the Internet. Few if any point out that the markets had already passed from the phase of reasonable valuations of investments to merely buying because it was going up – in short, a bubble.

> It is noteworthy that following every market movement, whether great or small, analysts always "explain" why these movements occurred – after the fact. But analysts are utterly incapable of predicting future movements. Thus, they explain everything and predict nothing.

Carlson did not discuss the possibility that the 170% gain of the S&P 500 Index in a mere five years might have represented a wild speculative bubble that had to be punctured sooner or later. The simplest explanation is that after the run-up from 1982 to 1987, stocks were going up mainly because they were going up. When they started down, they were going down because they were going down. Any pretense to "investment" had long since been discarded and herd behavior prevailed.

From Wednesday, October 14 to Friday October 16, the S&P 500 Index dropped from about 310 to about 283, a drop of about 9% in three days. Most of this decrease took place on Friday. On Friday, when futures became cheap compared to stocks, index arbitragers sold stocks and bought futures. Many institutions with program trading policies had been laggard in executing their programs, and whereas their models indicated they should have sold $12 billion of stocks, they had only sold about $4 billion by the close on Friday.[61] Thus, by the time of the opening on Monday morning, there was a substantial amount of pent-up selling pressure. Thus the market dropped precipitously on Monday morning, and by midday on Monday, October 19th, the S&P 500 Index was down to 255 and it closed the day at 225, a drop of about 20% in one day. As Carlson reported:

The record trading volume on October 19 overwhelmed many systems. On the NYSE, for example, trade executions were reported more than an hour late, which reportedly caused confusion among traders. Investors did not know whether limit orders had been executed or whether new limits needed to be set. Selling on Monday was reportedly highly concentrated. The top ten sellers accounted for 50 percent of non-market-maker volume in the futures market; many of these institutions were providers of portfolio insurance. One large institution started selling large blocks of stock around 10:00 in the morning and sold thirteen installments of just under $100 million each for a total of $1.1 billion during the day.

Carlson discussed contributing factors to the precipitous drop in terms of (a) program trading, (b) stock futures arbitraging, (c) inability of the financial system to provide timely information, leading to great confusion among investors as to the state of the market and their sell orders, leading to herd behavior, and (d) margin calls. He also mentioned that major banks extended credit on October 19, when strictly speaking, margin calls would have normally "sold out" investors' holdings, otherwise the debacle could have been worse.

Carlson described the actions of the Federal Reserve System to the market crash:

The Federal Reserve was active in providing highly visible liquidity support in an effort to bolster market functioning. In particular, the Federal Reserve eased short-term credit conditions by conducting more expansive open market operations at earlier-than-usual times, issuing public statements affirming its commitment to providing liquidity, and temporarily liberalizing the rules governing the lending of Treasury securities from its portfolio. The liquidity support was important by itself, but the public nature of the activities likely helped supsport market confidence. The Federal Reserve also encouraged the commercial banking system to extend liquidity support to other financial market participants. The response of the Federal Reserve was well received and was seen as important in helping financial markets return to more normal functioning.

Carlson also said:

In an effort to restrain the declines in financial markets and to prevent any spillovers to the real economy, the Federal Reserve acted to provide liquidity to the financial system and did so in a public manner that was aimed at supporting market confidence. One of the most prominent actions of the Federal Reserve was to issue a statement on Tuesday indicating that it would support market liquidity. This statement was referred to by one market participant as the most calming thing that was said Tuesday, and likely contributed to the rebound that morning. (*emphasis* added)

Here is a clear admission by the Fed that it acted to *"restrain the declines in financial markets."*

The S&P 500 Index gradually recovered somewhat and by the close on October 21, reached 255, a gain of about 13% from the close on October 19. That still represented more than doubling of the Index from five years earlier.

Donald MacKenzie[62] wrote a lengthy discourse on the crash of 1987. He particularly discussed the putative role of *portfolio insurance* in contributing to the crash. Various techniques are available to prevent major losses due to a downturn in securities prices. Each of these comes at a cost; one is willing to pay a definite small fee to prevent an improbable large loss. One approach is to sell a put option on the security one holds long. For example, suppose one buys 100 shares at 50, and sells a put option (at a cost of $500) for the right to sell 100 shares at 48. The cost of the long investment is $5,000, and the maximum possible loss is $700 ($200 for the stock price and $500 for the put option) regardless of how low the stock price drops. Portfolio insurance is the use of programmed trading whereby decisions are made to buy or sell securities based on the desire to prevent significant losses in a portfolio of investments. However, put options are unsuitable for large institutional investors. Another option is the stop-loss order that required that a stock be sold at the market if the stock price dropped to a pre-assigned level. Thus in the example above one could buy the stock at 50 and require that the stock be sold if the stock price touches 45. Portfolio insurance was a sophisticated programmed approach to shifting between stocks and cash (or government bonds) as stock prices fluctuated, buying stocks as prices rose and selling them as the value of the portfolio fell toward its floor. It was in a sense, the large institutional investor's equivalent of put options. During the 1980s a growing number of institutional investors began using portfolio insurance, many of them utilizing an algorithm supplied by the same consulting company. This algorithm involved arbitrage between stock holdings and futures contracts on the S&P 500 Index. It was understood by all involved that portfolio insurance would fail if a dreadful external event caused the market to fall discontinuously. If stock prices 'gapped' downwards for some reason, plunging discontinuously, there would not be sufficient time to adjust the portfolio accordingly. Furthermore, there were fears that if the use of portfolio insurance was widespread, a positive feedback effect might amplify price movements.

According to MacKenzie, in the 1980s

The demons of the 1970s – rampant inflation, oil shocks, trade union power – seemed to be receding, banished by liberalized markets, monetarism, Reaganism and the new breed of aggressive financial management, exemplified by the audacious junk bond acquisitions by asset stripping corporate raiders.

However, MacKenzie provided a sobering counter:

> By the autumn of 1987, however, doubts were growing as to whether the apparent successes of 'Reaganomics' were sustainable. The US trade deficit had ballooned, as had its public debt, the dollar was under pressure, and there were fears that interest rates would have to rise. On Wednesday October 14 disappointing US trade figures, and moves by the Ways and Means Committee of the House of Representatives to remove tax advantages that had contributed to the mergers and acquisitions boom led to what was then the largest ever number of points lost in a single day by the Dow Jones average.

This was a prelude to the cataclysmic drop of Monday, October 19, 1987. MacKenzie described some the activity of that day as follows:

> As alarming as the size of the crash were the breakdowns in markets that accompanied it. For prolonged periods on October 19 and October 20 the stocks of great US corporations such as IBM and General Motors – normally the most readily traded of all private securities – simply did not trade at all, as the New York Stock Exchange's specialists could not match buyers with sellers.
>
> Those who tried to sell via telephones often found they could not get through. Some brokers simply left their telephones to ring unanswered; others tried to respond but could not cope with the volume of calls.
>
> The trading disruptions in New York broke the link that arbitrage established between the stock and futures markets. If significant component stocks in the index were not trading, however, the calculated index value rapidly became stale; its relationship to market conditions became indeterminate. The breakdown in arbitrage permitted futures prices to plunge far below the theoretical values implied by the apparent level of the index. . . . The arbitrage that the discrepancy should have evoked was to buy futures and short sell the underlying stocks. . . . It was quite unclear [however] whether that arbitrage could successfully be completed.

In attempting to explain the crash, MacKenzie provided a lengthy analysis. A fundamental question was the degree to which portfolio insurance contributed to the crash of October 19. While many commentators have leapt to the immediate conclusion that portfolio insurance was a major factor in the crash, MacKenzie concluded that it was "immensely difficult to answer this conclusively."

As MacKenzie pointed out, on the one hand, portfolio insurers plus stop-loss orders accounted for a significant portion of stock and futures sales on October 19. On the other hand, only just over 1% of the US market's total capitalization was transacted during the crash, but that small percentage change in ownership was associated with a price decline of over 20%. MacKenzie also showed that there were widespread expectations for a market fall, since the price rise from 1982 to 1987 was reminiscent of 1924 to 1929, and there may have been an element of "self-fulfilling prophecy" involved (see Figure 1.3 in "Rationality of Experts?" in Chap. 1). In that connection, MacKenzie suggested that "it may be the rebound in the afternoon of October 20 and on October 21 that is more challenging to explain than the price declines on October 19, for which reasonably plausible explanations . . . can be found." He then made a very perceptive comment:

Sharp declines, not sharp rises, are regarded as undesirable and are thus in need of explanation.

The point is that when the stock market nearly tripled in 5 years, no explanation, justification or explication was required. However, when it dropped a mere 20%, all sorts of alarms were raised. The established viewpoint in the latter part of the 20th century was that it was right and natural for asset prices to rise enormously, and when they dropped, it must have been due to some sinister element that had to be investigated.

Overall, in this 70-page report, MacKenzie showed that the details of the 1987 crash are immensely complex and it is difficult to draw firm conclusions. MacKenzie did not seem to comment on the point that after the steep run-up in stock prices from 1982 to 1987, the transition had already passed from investing to mania, and much of the stock market activity was focused on buying stocks on upward momentum. Indeed the very essence of portfolio insurance is *momentum buying* when stocks go up, whereas *investment based on value* would buy stocks when they went down in the expectation that in the future the stock price of a good company would right itself. As the song goes, "A kiss is just a kiss . . .the fundamental things don't change." Bubbles are created by speculators, buying on momentum to sell to the next speculator. They eventually become vulnerable to puncture.

One of the remarkable things about the Crash of 1987, is not the crash itself, so much as the recovery from the crash. This was aided and abetted by the Federal Reserve, which pumped money into the banking system. As Greenspan said on the day after the crash:

The Federal Reserve System, consistent with its responsibilities as the nation's central bank, affirms today its readiness to serve as a source of liquidity to support the economic and financial system.

This is Greenspan-speak for asserting that the Fed would do everything it could to prop up asset bubbles.

THE DOT.COM MANIA

BOOM AND EUPHORIA

In the early 1990s, there was a wide expansion in the use of personal computers in business and at home. The US computer industry concentrated more upon computer software than hardware. Software produced large profits with high markups using minimum investment in facilities.

> Computer hardware became a commodity product, i.e. virtually indistinguishable from the product of any other competitor. Commodity products produce very little profits as each competitor constantly undercuts each other's prices. Asian companies, with small manufacturing costs, produced virtually all of the hardware components at this point. Software, however, was protected as intellectual property with patents. Therefore, a product such as Microsoft Windows is a one of a kind product. This creates a strong barrier to entry, a benefit that is highly sought after in business. The stock prices of software companies were marching ahead rapidly.
>
> Many small software companies were started by college students in garages,... Every startup wanted to become 'the Next Microsoft'. Eventually, several of these start-up companies took the notice of serious venture capitalists, who were looking to finance these operations, take them public and reap massive profits.... The majority of the software companies were started in Silicon Valley, near San Francisco....[63]

John Cassidy[64] described the origin of the Internet. The Internet concept grew out a Defense Department initiative (ARPA Net) originally developed for defense communications and information retrieval. In their presidential campaign, Al Gore and Bill Clinton advocated the "information superhighway." Cassidy said:

Almost immediately, businesses saw the internet as a profit-making opportunity. America Online made the Internet available for the masses. The Yahoo search engine was started in 1994. Amazon became the first online bookstore in 1994. EBay was started in 1995 as an online auction site. As the Internet moved from the hobbyist domain to a commercialized marketplace, online business owners became fantastically wealthy. Many technology companies were now selling stock in initial public offerings (IPO's). Most initial shareholders, including employees, became millionaires overnight. Companies continued to pay their employees in stock options, which profited greatly if the stock went up even slightly. By the late 1990's, even secretaries had option portfolios valued in the millions![65]

Cassidy described the events leading up to Netscape going public in 1995. The initial prospectus called for 3.5 million shares at about $13 for a valuation of about $450 million. Finally, 5 million shares were issued at $28 for a valuation of over a billion dollars. The stock rose as high as $74 on the first day of trading, and closed at $58, valuing the stock at almost $3 billion. Between 1992 and 1996 the market valuation of AOL stock rose from $70 million to $6.5 billion. Many economists claimed that we were in a *new economy*, where inflation was virtually nonexistent and stock market crashes were obsolete. It was claimed that earnings were no longer relevant in valuing stocks. New buzzwords like *paradigm shift* were prevalent. From 1996 to 2000, the NASDAQ Index increased from 600 to 5,000. Dot-com companies run by young entrepreneurs went public raising hundreds of millions of dollars in initial public offerings (IPOs). Most of these companies had no earnings and uncertain prospects. As Maggie Mahar said: [66]

> It didn't matter if the company was any good; if you downgraded it, you were almost certain to be wrong. And on Wall Street, the reality was that picking a good stock was far more important than picking a good company.

During the 1990s, Americans were pouring money into the stock markets, but predominantly into the dot.coms. The great expansion in 401(k) retirement plans, and the overwhelming preponderance of investment of those funds into stocks was an important factor. In a year and a half starting in 1995, the Dow-Jones average climbed 45% percent and the NASDAQ rose 65%. By the summer of 1996 there were 800,000 online stock trading accounts in the United States.

While this was happening, a few bears cautioned against the excesses of the bubble that was forming. Cassidy cited the views of two high-level managers at

Morgan-Stanley who said: "I believe that US stocks are overheated, overvalued, vulnerable to a cyclical bear market." The Morgan-Stanley bears went on to say:

> You've got stocks selling at absolutely unbelievable multiples of earnings and revenues. . . . You've got companies going public that don't even have earnings. You've got people setting up Internet pages to reinforce other's convictions in these highly speculative stocks. This is wild stuff out of the past. In every market where it has happened – from the US to Japan to Malaysia to Hong Kong – it always ends in the same way.[67]

They were right in principle, but wrong in practice. The markets still had a long way to go up before they came down. However, "sensible people" always seem to grossly underestimate the expansiveness of bubbles and are usually several years and several thousand index points early in their predictions of collapse. The extent of human greed during the euphoric phase is difficult to fathom for those that do not get caught up in the maelstrom.

The rise of the Internet stocks is shown in Figure 2.4.

GREENSPAN AND THE ROLE OF THE FEDERAL RESERVE

Cassidy provided a detailed review of the effects of actions and inactions of the Fed on the dot.com bubble, and specifically the persona of Alan Greenspan, the head of the Fed. As usual, an almost religious belief in the Fed prevailed on Wall Street. The stock markets reacted with incredible sensitivity to each hint of a rate change by the Fed, putting enormous pressure on Greenspan, who sat in the hot seat.

According to Cassidy:

> Greenspan was far from convinced [in 1996] that a speculative bubble had developed. He was coming to believe that the economy's performance justified higher stock prices. In Greenspan's opinion, many of the old rules of thumb didn't seem to work anymore.

Furthermore Greenspan was well aware of the blame the Fed took for raising rates in 1929. He had poured money into the economy after the 1987 crash and revived the economy. Raising rates to inhibit the growing bubble seemed risky to his political future. In 1998, Greenspan continued his policy of not increasing interest rates while the Internet stocks soared. Amazon announced plans to add CDs to its site, and the stock went from 40 to 140 in a few weeks. Broadcast.com lost $6.5 million on revenues of $6.9 million. The IPO opened at $18 and closed at $63.

At the end of January 1998, during an appearance on Capitol Hill, a senator asked Greenspan how much of the Internet stock boom was based on sound fundamentals and how much was based on hype? Cassidy quoted Greenspan's replies as follows:

First of all, you wouldn't get 'hype' working if there weren't something fundamentally, potentially sound under it.

The size of that potential market is so huge that you have these pie-in-the-sky type of potentials for a lot of different [firms]. Undoubtedly, some of these small companies whose stock prices are going through the roof will succeed. And they may very well justify even higher prices. The vast majority are almost sure to fail. That's the way the markets tend to work in this regard.

There is something else going on here, though, which is a fascinating thing to watch. It is, for want of a better term, the 'lottery principle'. What lottery managers have known, for centuries is that you could get somebody to pay for a one-in-a-million shot more than the value of that chance. In other words, people pay more for a claim on a very big pay-off, and that's where the profits from lotteries have always come from. So there is a lottery premium built into the prices of Internet stocks.

But there is a root here for something far more fundamental – the stock market seeking out profitable ventures and directing capital to hopeful projects before the profits materialize. That's good for our system. And that, in, fact, with all of its hype and craziness, is something that at the end of the day, probably is more plus than minus.

However, as Cassidy put it: "The speculative mania was starting to spiral out of control." But Greenspan was concerned about the ongoing financial upheaval in East Asia where most of the countries entered deep recessions in 1998. Greenspan was worried that a rise in US interest rates might lead to severe repercussions in the world economy. Cassidy said:

The Asian crisis had placed Greenspan in an awkward position. After sitting on the fence for a couple of years in the debate about whether there was a speculative bubble, he had now concluded that what was happening on Wall Street did indeed, represent a bubble, at least in part. But he still didn't accept that it was the Fed's duty to burst the bubble, and his concerns about Asia reinforced this reluctance.

> At the height of the dot.com boom, several Nobel prizewinning economists heed and hawed as to whether there was a bubble, and if there was a bubble, how serious it was.

Some of Greenspan's colleagues in the Fed urged greater monetary restraint in 1998. Two members of the Federal Open Market Committee (FOMC) voted to raise interest rates immediately. In the end, the FOMC backed Greenspan's decision not to raise rates by a 10–2 vote, but several members had misgivings.

In his public statements, Greenspan was (as usual) obscure – perhaps purposely. He argued on the one hand that the markets would likely stabilize of their own accord; on the other hand, "firming actions on the part of the Federal Reserve may be necessary to ensure a track of expansion that is capable of being sustained." As Cassidy said: "This was typical Greenspan: hinting at higher interest rates, but hedging his bets."[68]

In the late summer of 1998, Wall Street interpreted the ambiguity of Greenspan's remarks to imply that higher interest rates were finally on the way. Believing in the supreme power of the Fed, the stock markets slumped sharply. The next meeting of the FOMC was scheduled for late August 1998. Cassidy believed that had the FOMC raised interest rates then,

> It is conceivable that the Internet stock boom would have come to an end then and there. More likely, several interest rate hikes would have been necessary to burst the bubble. Either way, the next two years would have looked very different.

As it turned out, international events prompted Greenspan to hold off from raising interest rates, and the parade of Internet IPOs continued. The Russian government devalued the ruble and reneged on some of its debts. The Russian devaluation sparked an international financial crisis. As Cassidy described it:

> All around the world, financial markets shuddered, stabilized, then shuddered again. On Monday, August 31, 1998 the Dow fell by 513 points – its second–biggest points drop. The NASDAQ dropped 140.43 points – its biggest points fall. Internet stocks were particularly hard hit. Excite and Amazon.com both fell by more than 20 percent, Yahoo! and America Online by about 15 percent.... At the week's end, Time published a cover showing investors falling off a cliff-shaped stock chart, with the headline: "IS THE BOOM OVER?" With Asia already in a slump, Russia in turmoil, – and Latin America teetering, there were widespread fears of a global depression.

The biggest victims were hedge funds. Cassidy cites George Soros' Quantum Fund, which lost $2 billion in a few weeks, and Long-Term Capital Management (LTCM), which lost a similar amount (see "Long-Term Capital Management" in this chapter).

After being pressed for months to raise interest rates, Greenspan now found himself being urged to cut them in order to calm the markets. On September 4, 1998, Greenspan hinted at a future rate cut. But financial markets remained in chaos. On September 20, 1998, LTCM indicated that it was facing bankruptcy "and might have to unwind tens of billions of dollars' worth of investments." At the end of September 1998, the FOMC reduced the federal funds rate by 0.25%. But Wall Street had hoped for a bigger cut, and the markets remained in turmoil. Internet stocks slumped. By the second week of October, Amazon.com and America Online were both 40 percent off their highs. In mid-October, in a highly unusual move, Greenspan decided, on his own, to cut the rate another 0.25%. This time the markets reacted favorably. Without actually saying so, Greenspan had indicated his determination to prop up the stock markets.

Cassidy pointed out that the financial troubles of billionaires' hedge funds had little to do with ordinary Americans. Although the losses in the US stock market were more widely shared, the market averages were still far above where they had been a few years ago. In easing policy in such circumstances, Greenspan seemed to indicate that his policy was to prevent falling asset markets. Cassidy's assessment was:

> His reversal added to the growing belief that the Fed would always be there to bail out investors if anything went wrong, and this made investors even more willing to take risks.
>
> Greenspan, however inadvertently, ended up further inflating the Internet bubble. The two interest rate reductions confirmed to many people on Wall Street that in a crisis the Fed chairman could be relied upon to take prompt and dramatic action to protect their interests. Bill Dudley, the chief economist at Goldman Sachs, commented after the second rate cut: 'This is a way of telling everyone, the lifeguard is back on duty; you can go back in the pool'.

And so, the stock markets took off in even greater euphoria than before. As Figure 2.4 shows, the Internet stock index rose by almost a factor of ten from October 1998 to early spring of 2000. In 1999, there were 546 IPOs that raised over $69 billion. The average first-day gains of IPOs in 1999 were 68 percent compared to 23 percent in 1998.

As Cassidy pointed out:

Low interest rates weren't Greenspan's only contribution to the stock market boom. His frequent references to the benefits of new technology, and his refusal to criticize excessive speculation, also played an important role. In August 1999, Greenspan said stock prices reflected 'Judgments, of millions of investors, many of whom are highly knowledgeable about the prospects for the specific companies that make up our broad stock price indexes'. Instead of second-guessing these educated judgments, the Fed ought to stick to monitoring inflation pressures in the economy, he concluded.

Evidently, Greenspan was espousing the *intelligent market* doctrine, but history shows that markets are often ruled by greed and herd behavior, not intelligence. Cassidy argued that it is the Fed's responsibility to restrain such behavior to avoid the boom-bust cycles that existed so often before the Fed was created. Cassidy also pointed out that the Fed had other tools at its disposal such as raising the margin rate, but Greenspan refused to do this.

Cassidy asserted however, that Greenspan became increasingly worried about the stock market in late 1999. Since his upbeat Congressional testimony in August 1999, the NASDAQ had risen another 1,000 points, and speculative trading was rife. Cassidy suggested that privately, Greenspan "joked that he would like to introduce a law prohibiting day traders from buying a company's stock unless they could identify the product it produced." Nevertheless, according to Cassidy, as 1999 waned, it slowly began to dawn on Greenspan that a bubble had formed.

According to Cassidy:

Greenspan had been proceeding on the assumption that the Fed could concentrate on the real economy – inflation, unemployment, and productivity growth – and ignore the ups and downs of the stock market.

The "wealth effect" whereby people who made a good deal of money in the stock market felt wealthier and became free spenders, thus creating demand which in turn, heated the economy. Hence the stock market was not insulated from the economy, but became its principal driver. Industry began to expand capacity to meet the increased demand. However, Greenspan's belief was that this increase in capacity took time to come on line, whereas the rise in stock prices, and the resultant increase in consumer spending were immediate. Consequently, Greenspan was concerned that as overall demand in the economy rose faster than overall supply, inflationary pressures would build. While demand had been met temporarily by increasing the workforce and increasing imported goods, these buffers were used up, and price increases would result next.

Cassidy said:

This was a convoluted argument, which attracted criticism from academic economists ... but its internal logic mattered less than its practical consequences.[69] Greenspan had finally come up with an economic rationale for interfering with the stock market. To reduce the risk of inflation, the wealth effect would have to be attenuated. This 'does not mean that prices of assets cannot keep rising', Greenspan explained[70], 'only that they rise no more than income'. With personal income growing at about 6% a year this implied that stock prices could grow by 6% too. In the current environment, such an annual return was piddling. The NASDAQ had just returned almost 90 percent - in 1999 [and the Internet index was up 300%]. Investors weren't [necessarily] expecting a repeat performance in 2000, but they were looking for lot more than 6%. If Greenspan was serious about disappointing them, which he seemed to be, it could only mean one thing: higher interest rates were on the way.

On February 2, 2000, the FOMC announced a 0.25% rise in the federal funds rate. Cassidy pointed out that the language used to justify this rate increase was nearly the same as that used previously to justify no increase. Three weeks later, Greenspan appeared before the Senate Banking Committee, where he took a good deal of heat. The problem was that he was protecting against a putative inflation that had not yet shown up.[71] Greenspan repeated his argument from his January talk about the growing disparity between demand and supply but the senators were not convinced. One senator called the rate rise misguided and said the Fed's decision to raise interest rates was "more of a threat to our economy than inflation will ever be." As the NASDAQ approached 5,000, Senator Phil Gramm[72] suggested that equities were "not only not overvalued but may still be undervalued." It appears that the prevailing view in the Senate was that wealth not only could, but should be created by bidding up paper assets, and that it was right, natural and appropriate for stock market indices to double, double again, double again and keep on doubling until almost everyone was rich. The last thing they wanted in an election year was a stock market crash.

For the moment, technology investors continued to ignore the Fed and Internet stocks plowed higher ground, although "old economy" stocks weakened and the Dow-Jones average slipped.

BURSTING OF THE BUBBLE

By early March 2000, the stock markets had split. The "old economy" stocks in the Dow-Jones average were down but the NASDAQ, and especially the Internet stocks were still going strong. Since the start of 2000, more than 80% of the stocks comprising the S&P 500 Index were down 20% or more. But the technology sector was hitting new highs. The NASDAQ reached the astounding figure of 5,000 on March 10, 2000.

One disturbing aspect was that there was a very large increase in volatility of the stock markets in 2000, with some very big days, both up and down. The Dow gained 819 points on one of the up days. By early March there had been 15 days with daily changes of 3% or more in the NASDAQ (9 down and 6 up). Cassidy described trading on March 13 when the NASDAQ was down about 10% during the morning before it recovered in the afternoon.

Many of the fledgling Internet companies were short of cash. Companies raised cash with initial public offerings (IPOs) at a moderate price, and the great increases in stock price that took place afterward produced profits for investors and speculators, but not the companies themselves. For example "The Globe" raised $27 million in its IPO, but subsequent stock price increases raised the market valuation to $300 million. But that additional $273 million in capital gains was not available to the company coffers. Since most of these new Internet companies were essentially starting from scratch with almost no initial endowment, their needs for cash were great. Furthermore, many of the originators of these companies were young, inexperienced, and often replete with non-performing relatives and hangers-on. The "burn rate" at which they were spending money for start-up was alarming. A Barron's article in mid-March 2000 compared the cash burn rate with cash available for a large number of Internet companies and the result showed that many of them would run out of cash very soon. While they could theoretically raise more money via another offering of stock, the public did not seem to be in a mood to support such offerings from companies they had previously bought with enthusiasm, that were now floundering. On March 20, 2000, many of the stocks listed in the Barron's article tanked.

On March 21, 2000, Greenspan and the FOMC raised the federal funds rate by ¼ point. Despite that, the Internet stocks rallied once more. This was the last gasp of the Internet stocks. Cisco Systems passed Microsoft to become the highest valued corporation at $555 billion. Following this, several negative commentaries on Internet companies were published – there had always been such reviews – but this time for some reason, people seemed to pay attention. Between March 28,

2000 and April 3, 2000, the Internet stock index dropped 13.5% and now stood 35% below its value on March 10, 2000. Margin calls began to add to the selling exodus. The NASDAQ went through wild gyrations – up and down – in early April, but around April 10, 2000, serious selling resumed. From April 10 to April 13, 2000, the NASDAQ dropped 19% and the Internet index dropped 32%.

The debacle continued through 2000 and 2001. Many Internet stocks collapsed and never recovered. Two stock histories of Internet darlings are shown in Figs. 2.5 and 2.6. There are many similar examples that could be shown. *Yet Mr. Greenspan and top economists weren't sure if there was a bubble, or if there was a bubble, how serious it was!*

MERRILL-LYNCH IS BULLISH ON AMERICA

Merrill-Lynch had lagged behind other investment firms that had wildly and enthusiastically advocated dot.com stocks.[73] Toward the end of the dot.com boom, Merrill-Lynch was playing "catch-up" using their "expert" Henry Blodget as "point man."[74]

As the dot.com craze was reaching a feverish high in 2000, Merrill-Lynch Launched the *Internet Strategies Fund,* raising $1.1 Billion in an IPO on March 27, 2000, just before the crash of the dot.com bubble. As the markets crashed, so did the Fund:

> On October 5, 2001, the Internet Strategies Fund ceased to exist. From the much-hyped beginning in the spring of 2000 to the quiet demise in the fall of 2001, investors accrued losses of 81 percent, representing nearly $900 million. For the disservice provided to investors, Merrill Lynch collected fees of approximately $45 million.[75]

This led to several lawsuits to be brought against Merrill-Lynch in 2002. The main complaint alleged that the defendants engaged in a scheme that was intended to use Mr. Blodget's strong reputation and bullish ratings on Internet stocks to market the Internet Strategies Fund to unsuspecting investors. Over one billion dollars were invested in the Internet Strategies Fund by investors. The complaint alleged that defendants failed to disclose that: [1] at the same time that Blodget was recommending Internet stocks, he held unpublished negative views regarding those same stocks, [2] considerable conflicts of interest existed within Merrill Lynch which compromised the objectivity of Merrill Lynch Internet analysts, and [3] Blodget's favorable ratings on Internet companies were influenced by Merrill Lynch's desire to generate investment banking fees.

In one filing in 2003, the judge said:

> This case is yet another of the class actions following the long boom and eventual bust of the Internet sector of the securities markets. After years of unrestrained speculation in volatile and highly untested common stocks, the Internet bubble burst in the spring of 2000, dragging the prices of common stocks down with it, and generating a wave of litigation.[76]

Note that this judge was totally unsympathetic to the plaintiffs. The judge said:

> Plaintiffs' Amended Complaint here alleges nothing new, and their Opposition merely attempts to reargue the grounds for this Court's decisions in the Global Technology Fund, ... Absent any change in the applicable law — that is, absent the creation of any SEC regulation or other legal authority that would require a mutual fund to disclose the information Plaintiffs demand — the reasoning in these earlier decisions applies to the claims here.

Unbelievably – or believably if you are cynical – the judge ruled that Merrill-Lynch had no legal requirement to disclose the fact that they sold a billion dollars worth of securities with which to purchase stocks on which they privately held negative views.

Apparently the various lawsuits were combined and they dragged out until 2007 when they were finally settled on February 1, 2007:

> Merrill Lynch & Co. won approval Wednesday of a $40.3-million settlement of three lawsuits over claims it provided misleading analyst research about Internet companies. US District Judge John Keenan in New York approved the deal reached after investors appealed the 2003 dismissal of two of the cases. Keenan also awarded $9 million to lawyers who represented almost 400,000 investors who sued. Investors won 6.25% of the $645 million in damages they sought, which Keenan said was 'at the higher end' of the percentage of recoveries in class - action securities suits. The lawsuits were brought on behalf of shareholders in three Merrill mutual funds: the Internet Strategies Fund, the Global Technology Fund and the Focus Twenty Fund. The firm issued falsely optimistic research reports, and fund prospectuses failed to disclose investments in companies with which Merrill sought banking business, the investors claimed.

Merrill was named in dozens of investor lawsuits in 2002 after the firm issued what the investors said were misleading research reports about Internet companies. US District Judge Milton Pollack, who died in 2004, dismissed many of the actions, saying the individuals who sued were 'high-risk speculators' who wanted to 'twist the securities laws into a scheme of cost-free speculators' insurance'. An appeals court upheld most of the dismissals. In February 2006, Merrill paid $164 million to settle 12 cases pending in the trial court and 11 on appeal.[77]

This case is remarkable because it seems evident that Merrill Lynch did not reveal to 400,000 investors their internal beliefs about dot.com stocks, spent many millions in legal fees to oppose plaintiffs' claims, and with the aid, support and endorsement of the courts, succeeded in limiting their responsibility to a pittance.

OTHER BUBBLES AND SWINDLES OF THE LATE 1990s AND 2000s

According to K&A, the major impact on the United States in the 1990s was the "revolution in information technology and new and lower-cost forms of communication and control that involved the computer, wireless communication and e-mail." We have already seen in "The Dot.Com Mania" in Chap. 2 that the dot.com bubble was of gigantic proportions and the levels of mass insanity reached in early 2000 far exceeded anything experienced in 1929.

In the first part of this section, we review a few specific instances of bubble mania and swindles from that era in greater detail: Adelphia, rogue traders at banks, and Enron. We also review Long-Term Capital Management, which was not a swindle, but it provides insights into the bubble mentality that prevailed, as well as the attitude of the Federal Reserve toward propping up markets. The Ponzi schemes of Albania provide an example of a rather incredible bubble in our time.

In the second part of this section, we note the widespread collusion that took place in the 1990s between corporations and the major accounting firms in misrepresenting data to enhance stock prices. Here, we provide capsule coverage of a selection of company accountants conspiracies that were heavily fined for illegal actions. Cheating, misrepresentation, stealing and fraud have been rampant in American corporations and the major accountancy firms have played a major role in these frauds. In an environment of interpreting "deregulation" as "no regulation" with laxity from all Government regulating agencies, such behavior was bound to expand.

ADELPHIA

Adelphia was a rather average dot.com bubble company that achieved its main notoriety through fraud. Its stock peaked at $84/share in 1999, and became worthless when it declared bankruptcy in mid-2002. From 1998 through March 2002, Adelphia, the nation's sixth largest cable-television company, systematically and fraudulently excluded billions of dollars in liabilities from its consolidated financial statements by hiding them on the books of off-balance sheet affiliates. It also inflated earnings to meet Wall Street's expectations, falsified operations statistics, and concealed blatant self-dealing by the Rigas family that founded and controlled Adelphia.

A footnote to an earnings release in 2002 revealed that $2.3 billion of off-balance-sheet debt had been incurred through co-borrowings by the Rigases. The loans and other related-party transactions became the object of SEC scrutiny and grand jury investigations. Later in 2002, a federal grand jury in Manhattan indicted five former Adelphia executives on 24 counts of securities fraud, wire fraud, and bank fraud. Their actions were described by an US Attorney as "one of the most elaborate and extensive corporate frauds in history."[78]

As CFO magazine described it:

It was a shocking end to the ruling family's hold on an empire that was built over 50 years with the purchase of a tiny cable franchise for $300 and a $40,000 loan. . . . Rigas never wavered from his extremely centralized management style. It was still being run as if it were a small family business. . . . The Rigases seemingly ran the company as if it were their own private cash machine. The family has been accused of commingling the accounts of Adelphia with their other businesses, borrowing–and at times allegedly stealing–to pay for lavish homes and other personal expenses, including a private jet and construction of a golf course.

The cases were prosecuted in the courts for several years and in June 2005, John Rigas was sentenced to 15 years in prison, and his son, Timothy Rigas was sentenced to 20 years.

During the time the Rigas family ruled (and milked) Adelphia, like Keating in the S&L business, they ran a continual anti-porn and anti-smut campaign. Both the Rigas and Keating apparently wanted nothing improper in the bedroom but did not apply the same scruples to the boardroom. It is ironic that the residual Adelphia Company that derived from the outcome of the bankruptcy proceedings became the nation's only leading cable operator to offer the most explicit category of hard-core porn.

ROGUE TRADERS AT BANKS

At various intervals, rouge traders in comparatively low positions at major banks invest large sums of money in speculative schemes that end up in financial disaster at a large scale. While technically, these are frauds rather than bubbles, it is the bubble atmosphere in markets that provides the environment for these rogue traders to go undetected until they have lost billions.

Three highly publicized cased of rogue traders losing billions for banks in illicit speculative trades were:

- Nick Leeson, who lost $827 million causing collapse of Barings Bank in England in 1995.

- John Rusnak who lost $691 million in 2001 at Allied Irish Banks – Ireland's second biggest bank.

- Jerome Kerviel lost over $7 billion at Societe Generale in France in 2007.

There seems to be a six-year period between such scandals. In late 2008 it was revealed that Bernard Madoff's hedge fund was actually a Ponzi scheme to the tune of $50 billion, dwarfing previous frauds.

BARINGS BANK

Nick Leeson was a Londoner who worked for Barings in their Singapore office. Leeson and his traders were authorized to transact futures and options orders for clients and arbitraged price differences between Nikkei futures traded on the SIMEX and Japan's Osaka exchange.

These did not seem like risky investments:

However, Leeson took unauthorized speculative positions primarily in futures linked to the Nikkei 225 and Japanese government bonds (JGB) as well as options on the Nikkei. He hid his trading in an unused account, number 88888. Exactly why Leeson was speculating is unclear. He claimed that he originally used the 88888 account to hide some embarrassing losses resulting from mistakes made by his traders. However, Leeson started actively trading in the 88888 account almost as soon as he arrived in Singapore. The sheer volume of his trading suggests a simple desire to speculate. He lost money from the beginning. Increasing his bets only made him lose more money.... On February 23, 1995, he hopped on a plane to Kuala Lumpur leaving behind a $827 million hole in the Barings balance sheet.

What is amazing about Leeson's activities is the fact that he was able to accumulate such staggering losses without Barings' management noticing.... By falsifying accounts and making various misrepresentations, he was able to secure funding from various companies within the Barings organization and from client accounts.... Leeson was an accomplished liar. He falsified records, fabricated letters and made up elaborate stories.[79]

Some of his methods were amazingly primitive. For example, he cut and pasted old letterheads to create bogus confirmation faxes and used them to reassure the London head office that there was money on its way to balance out the losses he had incurred. Barings management was blissfully unaware of his shenanigans:

Six days after fleeing Singapore, Leeson was arrested [and] returned to Singapore to stand trial. Convicted of fraud, he was sentenced to six and a half years in Singapore's Changi prison.... For good behavior, he was released from prison early in July 1999.

ALLIED IRISH BANKS

John Rusnak worked for Allied Irish Banks, and had a need to recoup money he had lost on a proprietary trading strategy sometime in 1997. He later compounded the situation by speculating, racking up huge unrecorded liabilities for the bank. This was discovered in 2001 and the total loss from the debacle was estimated to be $691 million. This wiped out more than half of the bank's 2001 earnings and weakened the financial position of the bank. As in the case of Leeson, Rusnak used a variety of techniques to cover up his losses.[80]

SOCIETE GENERALE

Jerome Kerviel's case is an odd one in some respects. His motivations are mysterious, because it appears that he made no personal gain from the unauthorized trades. It is amazing that this 31-year-old trader was dealing with more than $73.3

billion – a sum that was greater than the bank's market capitalization of $52.6 billion. According to one Internet source:

> Kerviel had been betting throughout 2007 that markets would fall – a winning position. But the trader overstepped his authority and wagered much more money than he should have. So at the beginning of January 2008, Kerviel voluntarily created losing positions to neutralize his earlier gains and cover his tracks. But the steep drop in the markets in 2008 expanded these losses far beyond what he expected. Had he maintained his negative stance, he would be even further ahead. The bottom line seems to be a loss of more than $7 billion.[81]

ENRON

Public utilities have the task of providing power, gas, communications, water and other needs to the public. In the United States, the regulation of energy utilities dates back to the 1930s, when the Public Utility Holding Company Act of 1935 was passed because the utility holding companies were believed to have been major contributors to the 1929 stock market crash. Since the 1930s, the prevailing view was that utilities must be regulated to (1) assure that the public is properly served at reasonable rates, and (2) the utility is entitled to a fair profit. This system worked very well and there never was any need to change it. The utilities had to be run primarily for the benefit of the public they served, but they also had to be allowed to earn a reasonable profit.

However, starting with the Reagan administration (1980–1988) and continuing to the present day, it became fashionable in Washington and university economics departments to believe that deregulation of utility markets, indeed deregulation of just about everything, would make everything more efficient. The Reagan view was based on an almost religious antagonism to any form of government regulation of business. In this system, the utilities are run for the benefit of their stockholders, or worse still in some cases (e.g. Enron) for the benefit of the management – and often to the detriment of the public.

The Enron story is not unlike the S&L story (see "The Savings and Loan Scandal of the 1980s" in this chapter) in that, once deregulation was legislated, shrewd, amoral market manipulators turned what used to be a public service into an illegal program to build their own personal wealth. Once again, deregulation has been shown to be detrimental to the public.

Enron was originally a rather humdrum natural gas utility not worthy of any special attention. With the advent of deregulation under the Reagan administration, it acquired additional utilities, changed its name to Enron, and entered a new phase of its endeavors under leadership of newly appointed CEO, Kenneth Lay. Enron owned a large network of natural gas pipelines across the United States. These provided the cash flow that enabled other ventures and investments. They were the only part of Enron that made significant operating profits. In these other ventures, the approach used by Enron was intended to (1) spend great sums of money to influence regional legislators to pass deregulation policies favorable to Enron, (2) buy up control of suppliers of utilities in these regions where deregulation was in force, (3) use their control of the regional utility supplies to force up prices paid by suppliers to end-users (i.e. the public) and thereby make huge profits at the public's expense.

California Electric Power:

The best example of Enron's operations is their cornering of the electricity market in California in the 1990s.[82] California was the first state to deregulate its energy markets. Until then, three investor-owned utilities served the state. The prevailing belief in the California Public Utility Commission was that if California opened up its electricity market to competition, the state's utility bills would drop significantly.

In June 1994, Enron vice president Jeffrey Skilling testified to the California State Commission that the State could save $8.9 billion a year by deregulating. In September 1995, Enron and several other companies submitted their deregulation plan to California policy makers. The California utilities lobbied in favor of this plan, and were unable to foresee that it would bring about their own downfall. In the fall of 1996, the California Legislature essentially was conned into using that plan as the basis for its energy deregulation bill. While the California legislature has a long history of poor judgment, adoption of the power deregulation plan stands out as one of the worst decisions of all time.

Incredibly, the deregulation plan called for the investor owned utilities (IOUs) – primarily Pacific Gas and Electric, Southern California Edison, and San Diego Gas and Electric – to sell off a significant part of their power generation to wholly private, unregulated companies such as AES, Reliant, and Enron. The buyers of those power plants then became the wholesalers from which the IOUs needed to buy the electricity that they formerly produced themselves. While the selling of power plants to private companies was part of "deregulation," the California legislature naively expected that there would be regulation by the FERC that would prevent market manipulation. The job of the FERC, in theory, is to regulate and enforce Federal law, preventing market manipulation and price manipulation of energy markets. When called upon to regulate the out-of-state privateers that were clearly manipulating the California energy market, the FERC hardly reacted at all and did not take serious action against Enron, Reliant, or any other privateers. The resources of the FERC are in fact quite sparse in comparison to their entrusted task of policing the energy market. In addition, lobbying by private companies clearly slowed down regulation and enforcement.[83]

Sharp's article[84] provides quotations by some of the architects of California's deregulation bill in the State Legislature:

We didn't foresee the problems, Shame on us for not passing a better law.

According to Sharp, one thing that went wrong in California was that the wholesale market was deregulated, but the retail side for consumers was still regulated. As a result, utilities were buying power at very high prices in wholesale markets and selling at low prices to their retail clients.

Another problem was that utilities were forbidden by law to enter into long-term contracts. Instead, they had to buy power on the spot market.

When the spot market became volatile, the utilities had to pay exorbitant prices. A third problem was that energy traders such as Enron cornered the electricity supply market and were withholding supplies, forcing prices to high levels. The wholesale price of electricity had climbed from $20 per MWH at the start of deregulation to $250 per MWH in 1999, even though demand had been relatively flat. The state's utility operator wanted to cap wholesale prices:

But Enron and the other suppliers threatened to take their power elsewhere. By late 1999, wholesale prices exploded to $750 per MWH. California declared that price gouging was widespread and capped prices. This angered the private companies, including Enron chairman Kenneth Lay. He wrote to the FERC, urging it to nullify the price caps. On Nov. 1, 2000, the agency removed the caps. At the height of the state's power crisis, the price of electricity boomed to $3,000 per MWH. Some household utility bills were $800 a month – more than rent. All told, the state's cost rose from $6 billion in 1999 to $27 billion in 2000 and $27 billion in 2001. . . .[85]

As the utilities were finding it more and more difficult to afford to provide electric power to customers in California, rolling blackouts became commonplace. The utilities asked for the right to pass on their higher costs to end-users. The public Utilities Commission was slow to react. The Governor of California, Gray Davis and the legislature waffled and wavered. As blackouts increased and PG&E declared for bankruptcy, the State finally began to act in 2001. The State took over the electrical power business and entered into long-term power contracts that excluded Enron. By late summer of 2001, power costs were down to $100/MWH.

Ultimately, the cost of the debacle has been estimated to be as high as "$71 billion, reflecting the cost to California consumers in overcharges, bailouts, and other associated costs."

In retrospect, it seems probable that had Enron acted with some restraint, and raised energy prices enough to make a decent profit, but not so high as to bankrupt the utilities and raise the hackles of the public, it might have prospered for a long time in California. But Enron's greed was insatiable and their arrogance was unbounded.[86]

World Operations:

The California electricity market was just one of many schemes that Enron was involved in. Almost all of these schemes involved Enron investing in recently deregulated energy and communications, typically by trading in futures contracts. With no upper limit to prices in these markets, the potential for profits was great. A great deal of press coverage has addressed the financial chicanery, lies and cover-ups of Enron as if that caused their demise, but in fact it was their financial failure that led to the misrepresentations and fraud in their reports. The big question is: "If these ventures were so profitable, why then did Enron fail?"

It appears likely that it is not so much that Enron soared and then collapsed, as much as it never succeeded much in the first place and used obfuscation and imaginative accounting to create the false impression of success for a few years.[87] McLean said:

> Start with arrogance. Add greed, deceit, and financial chicanery. What do you get? A company that wasn't what it was cracked up to be.... No one could explain how Enron actually made money.

In retrospect, it appears the Enron was basically a hedge fund trading in energy and communications futures, and was subject to the risks inherent in such operations. Its operating profit was modest and the relationship between cash flow from operations and reported earnings was difficult to perceive. Enron made a number of very bad investments on overseas projects – in India and Brazil for example. But the truth of their profitability remains difficult to decipher. Many of Enron's claimed assets and profits were inflated, or even wholly fraudulent and nonexistent. Debts and losses were put into entities formed "offshore" that were not included in the firm's financial statements, and other sophisticated and arcane financial transactions between Enron and related companies were used to take unprofitable entities off the company's books. Kenneth Lay seems to have been a great "snake oil salesman" who convinced Wall Street to support his stock with a very high price/earnings ratio based on very little hard evidence.

The following is abstracted from a Public Citizen's Report: [88]

Enron developed mutually beneficial relationships with federal regulators and lawmakers to support policies that significantly curtailed government oversight of their operations. Dr. Wendy Gramm, in her capacity as chairwoman of the Commodity Futures Trading Commission (CFTC), exempted Enron's trading of futures contracts in response to a request for such an action by Enron in 1992. At the time, Enron was a significant source of campaign financing for Wendy Gramm's husband, US Senator Phil Gramm. Six days after she provided Enron the exemption it asked for, Wendy Gramm resigned her position at the CFTC. Five weeks after her resignation, Enron appointed her to its Board of Directors, where she served on the Board's Audit Committee. Her service on the Audit Committee made her responsible for verifying Enron's accounting procedures and other detailed financial information not available to outside analysts or shareholders. Following Wendy Gramm's appointment to Enron's board, the company became a significant source of personal income for the Gramms. Enron paid her between $915,000 and $1.85 million in salary, attendance fees, stock option sales and dividends from 1993 to 2001. The value of Wendy Gramm's Enron stock options swelled . . . to as much as $500,000 by 2000. Phil Gramm was the second largest recipient in Congress of Enron campaign contributions, receiving $97,350 since 1989. . . . Enron spent $3.45 million in lobbying expenses in 1999 and 2000 to deregulate the trading of energy futures, among other issues. In December 2000, Phil Gramm helped muscle a bill through Congress without a committee hearing that deregulated energy commodity trading. This act allowed Enron to operate an unregulated power auction that quickly gained control over a significant share of California's electricity and natural gas market. Phil Gramm's legislation was in conflict with the explicit recommendations of the President's Working Group on Financial Markets, which is composed of representatives from the Department of Treasury, the Board of Governors of the Federal Reserve, the Securities and Exchange Commission and the Commodity Futures Trading Commission. The Working group expressly recommended against deregulating energy commodity trading because the traders would be in strong positions to manipulate prices and supply.

Investigations by state and federal officials concluded that power generators and power marketers intentionally withheld electricity, creating artificial shortages in order to increase the cost of power. Enron took advantage of lax oversight following deregulation and formed a complicated web of more than 2,800 subsidiaries — more than 30 percent (874) of which were located in officially designated offshore tax and bank havens. President Bush's presidential campaign received significant financial support from Enron ($1.14 million).[89]

Upon assuming office in 2001, Bush promptly scrapped plans put into place by former President Bill Clinton to significantly limit the effectiveness of these countries as tax and bank regulation havens. This action came at the height of high West Coast energy prices, probably allowing Enron to siphon billions to its offshore accounts. At the same time, the Bush administration and certain members of Congress waged a legislative and public relations campaign against the imposition of federal price controls in the Western electricity market. Such price controls remove the ability of companies exercising significant market share to price-gouge by effectively re-regulating the market. Bush's opposition to price controls unnecessarily extended the California energy crisis and cost the state billions of dollars. When federal regulators finally imposed strict, round-the-clock price controls over the entire Western electricity market on June 19, 2001, companies operating power auctions (like Enron) no longer had the ability to charge excessive prices and no longer had incentive to manipulate supply. While price controls clearly saved California, Enron suffered because it could no longer manipulate the market and price-gouge consumers. With no significant asset ownership to offset its losses, Enron's unregulated power auction quickly accumulated massive debts. At the same time, the curtailed revenue flow made it more difficult for executives and members of the Board to conceal the firm's accounting gimmicks. ... Due to Wendy Gramm's position on Enron's Audit Committee, she had intimate knowledge of Enron's financial structure and had access to sensitive financial information not available to Wall Street analysts or average shareholders. It is therefore probable that she knew of Enron's possibly fraudulent practices for some time and that her husband would have known as well. Enron's 874 tax haven subsidiaries allowed Enron to funnel billions of dollars to offshore accounts. The Gramms' close involvement with Enron's corporate and legislative activities, the Gramms' possible knowledge and/or connection to criminal misconduct relating to Enron's collapse, and the effects of Enron's layoffs and other economic impacts on Senator Gramm's constituents may have been the leading factor in Gramm's decision on September 4 not to seek re-election to the Senate in 2002.[90]

Clearly, deregulation is a policy favored by politicians who usually make profits from connections to companies that engage in deregulated machinations.

LONG-TERM CAPITAL MANAGEMENT

John Meriwether began his career in arbitrage at Salomon Brothers in 1977, running their Domestic Fixed Income Arbitrage Group. In 1993, Meriwether decided to begin Long-Term Capital Management (LTCM) with a number of mathematically oriented academics. In addition to his original team from Salomon, Meriwether recruited two financial superstars. He raised $2.5 billion based on his reputation. He also charged very high management fees, and required a three-year commitment for investments in the fund.

LTCM managed money for about 100 investors, it had about 200 employees, and it was in business for than five years. The fund had $140 billion in assets – all of which except for $2.5 billion were borrowed. This made it one of the largest investment funds.[91]

In the arbitrage game, the investor sells one security and buys another. If there is a slight advantage to the one that is bought, a profit is made, albeit a small one. However, if one can borrow large sums and leverage one's investment, significant profits can be made. Thus, if one has a leverage ratio:

{(amount invested in securities)/(amount invested by stockholders)}

of say, 25 (a typical operating point for LTCM), even a 1% profit on the investments yields a 25% profit for the stockholders. Achieving such a high leverage ratio requires that banks are willing to lend the additional funds. The reputations of Meriwether and his colleagues provided them with the basis to secure these funds.

While LTCM invested in a large number of arbitrage securities, two of their favorite types of investments were (1) arbitrage between government bonds and riskier corporate bonds, and (2) arbitrage between currencies from countries paying significantly different interest rates.

Corporate bonds, being riskier, pay a higher interest rate than government bonds. As long as the spread in yields between the two remains stable, going long on corporate bonds and short on government bonds yields a profit. To further increase the leverage, one may invest in corporate "junk bonds" with low rating to increase the yield even more. According to Lowenstein:

> The basic idea behind their trades lay in the *Efficient Market Hypothesis*. Over time, markets become more efficient, and the uncertainty associated with riskier assets decreases. Thus spreads between riskier and less risky assets should decrease.

If that were the case, price appreciation of the riskier asset should add to the profit. That is, the price of the riskier investment is initially low because of uncertainty in its future, but as its future becomes clarified, the price will appreciate.[92]

In the case of currency transactions, one of the favorite arbitrage situations in the 1990s was the so-called "yen carry" trade. Starting in 1995, there was strong economic growth in the United States while Japan was wallowing in the doldrums. Japan lowered its interest rates dramatically to try to stimulate its economy, and its currency weakened. The combination of an appreciating dollar and the large interest rate differential between Japan and the United States created a profitable trading opportunity based on borrowing yen, buying dollar assets, and gaining both on the appreciation of the dollar and the interest rate differential. This "yen carry" trade was widespread among hedge funds, trading desks of investment banks, and even some corporations. Japanese banks also resorted to the yen-carry trade by accumulating foreign assets.

The nature of arbitrage with highly leveraged investments is that profits are unpredictable from month to month, but if the theory is correct, monthly fluctuations should even out and add up to a preponderance of gains, resulting in yearly profits. The LTCM fund was immensely successful in its first four years of operation as shown in Table 2.3.

However, by 1997, the opportunities for investment in arbitrage had diminished, partly due to external events, and partly due to the entry of more players into the same ball field. Earnings were still good but far less than in previous years. LTCM decided to refund to investors about half of their original investments:

LTCM's very success bred many imitators in the proprietary trading desks of the major investment banks. As more and more players with similar trading strategies crowded into the market, the spreads narrowed on the favored convergence trades, eroding the profit margin for all the players. The relative tranquility of the markets also lulled the players into a false sense of security and spurred them on to increase their leverage, which reduced the spreads

Table 2.3 Average yearly earnings by LTCM before fees

Year	% Gain
1994	28%
1995	59%
1996	57%
1997	25%
August 1998	−45%

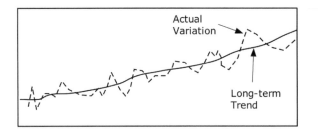

Figure 2.7 Variation of an economic parameter over time, comparing actual variations with the long-term trend.

further. By the spring of 1998, the convergence funds had to venture into new and uncharted markets in order to find profitable trades. The scene was set for a reversal of some kind.[93]

It was understood that all of these arbitrage operations were subject to risk, and some rather sophisticated mathematical risk-management models were employed. But ultimately, these were based on extrapolations of past relationships into the future using "normal" distributions. A simplistic description follows.

Suppose one has historical data on an economic parameter, as shown in Figure 2.7.

Now, if one plots the differences between the actual values and the long-term trend, one obtains Figure 2.8. If a so-called "normal" distribution equation is fitted to these data, it may be concluded that the probability of a deviation from the long-term trend at the far right or far left of this diagram is exponentially small. Hence the risk of such a large departure from the predicted trend is extremely small.

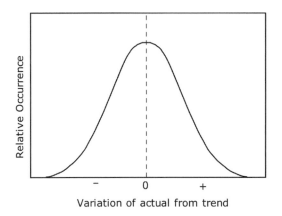

Figure 2.8 Relative occurrence of variations of actual data from the long-term tren.

According to LTCM models, the probability of a 10% loss in its portfolio was estimated to be an event that would occur once in a thousand (or so) trading periods. The probability of a loss of 50% in its portfolio was thought to be one-in-a-billion.[94] The problem with this argument is that there may be other critical factors, perhaps not obvious, that by luck or by policy, were constant during the period covered by Figures 2.7 and 2.8. If one or more of these factors were to change significantly in the future, a wide departure from this picture of stability could result.

Taleb wrote a detailed book explaining how this occurs in real life.[95] He provided a number of excellent examples of how extrapolation from the past can lead to an incorrect prediction of the future. One example is the growth of bacteria in a closed container with fixed amount of nutrients. Initially, the growth is exponential; however, eventually it maximizes and finally decreases sharply as nutrients are used up. Taleb also expounded at length on the example of a Thanksgiving turkey being groomed for Thanksgiving dinner in a few years. For a thousand days, all is well; the turkey is well fed and well taken care of. As far as the turkey knows, his life expectancy is infinite. Yet, his life is suddenly and unpredictably (from the turkey's point of view) terminated (see Figure 2.9).

Taleb showed that Figure 2.8 does not provide an accurate description of reality in most economic situations, and there are typically long "tails" that permit extremely large deviations with small, but non-negligible probability. Thus, Figure 2.8 should be replaced by Figure 2.10. These improbable cataclysmic events exert major forces on economic trends at various intervals. In Taleb's earlier book, "Fooled by Randomness," he emphasized the role of luck in

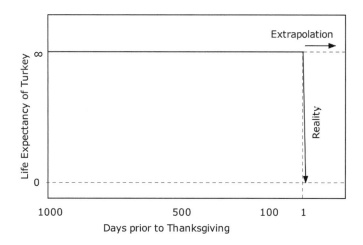

Figure 2.9 Life expectancy of a turkey (turkey's point of view).

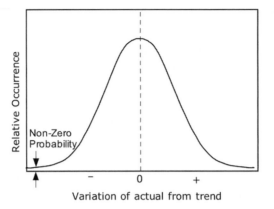

Figure 2.10 Relative occurrence of variations of actual data from the long-term trend with extended "tails".

investing. He provided many examples of commodity traders who had runs of luck making them wealthy in a short time, who believed their luck was skill. When bad luck and unpredictable events subsequently occurred, it wiped them out.

The summer of 1998 was a particularly turbulent episode for the financial markets of the United States and Europe. The following factors contributed to the reversal of fortune for the LTCM:

Salomon departs the field. The disbanding of the Salomon Brothers bond arbitrage desk on July 6th set in motion an internally reinforcing feedback process (see "Internal Feedback and Endogenous Risk" in Chap. 1). As long positions were sold, and short positions were bought back, other traders with similar positions were subjected to adverse price shocks. For traders with high leverage, this would trigger margin calls on their losing positions. They would be forced to unwind their trades, thus reinforcing the previous adverse price moves. Danielsson and Shin emphasized that:

> The unprecedented price moves in the summer of 1998 were not simply the result of extremely bad luck. Given the extensive copycat behavior of other traders and the large implicit leverage involved, it was only a matter of time before the system would be hit by a small outside shock that would send it into reverse. Once the system began to go into reverse, the internal dynamics of the feedback loop would take hold with a vengeance, and send it into a tailspin. The probability of this collapse is far from zero. Under the right conditions, it is a near certainty.[96]

Russia defaults. On August 17, 1998, Russia announced it was restructuring its bond payments—a de facto default. The losses forced many investment banks, hedge funds and other institutional investors to reduce their positions en masse. The resultant flight to higher quality investments boosted prices for Treasury bonds and sunk prices for lower quality bonds in an unprecedented fashion. Credit spreads had never moved so far so fast. LTCM's losses were huge because they were long on bonds and short on Treasuries. On August 21 alone, the firm lost $550 million. In late August, the fund found itself down 44 percent for the year. The models had judged that kind of loss to be something that occurs once in several billion times the life of the universe.[97]

Fall of the dollar against the yen. Danielsson and Shin emphasized that the dollar had been strong against the yen for several years, reaching a high of 147 yen per dollar on August 11, 1998. Many commentators were predicting that the yen/dollar ratio would reach 150 or perhaps 200 by the end of the year. However, in the aftermath of shocks to the international financial systems, the dollar fell to less than 120 yen by October 8, 1998. For those in the "yen carry" trade, who were long the dollar and short the yen, the result was disastrous. The effect of stop-loss orders contributed to the acceleration of decline of the dollar.

As losses mounted, LTCM had increasing difficulty meeting margin calls and needed more collateral to meet its obligations. The fund had great difficulty liquidating its positions and was now in very serious difficulty. On September 2, 1998, the partners sent a letter to investors acknowledging the fund's problems and seeking an injection of new capital to sustain it. That information soon leaked out and the fund's problems became common knowledge. LTCM's situation continued to deteriorate in September 1998, and the fund's management spent the next three weeks looking for assistance in an increasingly desperate effort to keep the fund afloat. However, no immediate help was forthcoming, and by September 19, 1998 the fund's capital was down to only $600 million, with an asset base of $80 billion so its leverage ratio was about 130—an indication of impending doom. It did not appear that LTCM could make it through the next week without outside assistance.[98]

Kevin Dowd's article dealt at length with the intervention of the Federal Reserve in the failing LTCM and whether such intervention was appropriate and needed:

Wall Street and the Federal Reserve had observed LTCM's deterioration with mounting concern. Many Wall Street firms had large stakes in LTCM, and there was also widespread concern about the potential impact on financial markets if LTCM were to fail. . . .

The LTCM fund partners persuaded a delegation from the New York Federal Reserve and the US Treasury that LTCM's situation was "much worse than market participants imagined." The Fed decided that some form of support operation should be prepared very rapidly to prevent LTCM's failure and thereby avoid what the Fed claimed they feared might otherwise be "disastrous effects on financial markets."

Accordingly, the New York Federal Reserve began to arrange a consortium of private companies to prepare a rescue package if no one else took over the fund in the meantime. In the interim:

> ... a group consisting of Warren Buffett's firm, Berkshire Hathaway, along with Goldman Sachs and American International Group, offered to buy out the shareholders for $250 million and put $3.75 billion into the fund as new capital. That offer would have put the fund on a much firmer financial basis and staved off failure. However, the existing shareholders would have lost everything except for the $250 million takeover payment, and the fund's managers would have been fired. The motivation behind this offer was strictly commercial; it had nothing to do with saving world financial markets.[99]

This group apparently felt that it could make a profit out of the corpse of LTCM if they could buy it cheap enough. The management of LTCM rejected this offer, presumably "because they were confident of getting a better deal from the Federal Reserve's consortium." The Fed therefore worked with its consortium to create an alternate rescue package:

> ... which was promptly accepted by LTCM and immediately made public. Under the terms of the deal, 14 prominent banks and brokerage houses— including UBS, Goldman Sachs, and Merrill Lynch but not the Federal Reserve—agreed to invest $3.65 billion of equity capital in LTCM in exchange for 90 percent of the firm's equity. Existing shareholders would therefore retain a 10 percent holding, valued at about $400 million. This offer was clearly better for the existing shareholders than was Buffett's offer. It was also better for the managers of LTCM, who would retain their jobs for the time being and earn management fees they would have lost had Buffett taken over. Control of the fund passed to a new steering committee made up of representatives from the consortium, and the announcement of the rescue ended concerns about LTCM's immediate future. By the end of the year, the fund was making profits again.[100]

Dowd then went on to ask: "Was the Federal Reserve justified"? The House Committee on Banking and Financial Services ran a hearing on the issue.

Among those testifying were the president of the New York Federal Reserve, William McDonough, and the chairman of the Federal Reserve Board, Alan Greenspan. Both McDonough and Greenspan defended their solution as:

> ... a private sector solution to a private-sector problem, involving an investment of new equity by Long-Term Capital's creditors and counterparties.

While some have claimed that the Federal Reserve had "bailed out" LTCM, they insisted that:

> No Federal Reserve official pressured anyone, and no promises were made. Not one penny of public money was spent or committed.

Dowd attacked these positions, arguing that there was no need for the US Government to be involved at all, and the Buffet team had found a solution without government interference. Dowd has a point here, but from a practical point of view, no harm was done by government action, and the end result seems to have made more sense than the Buffet proposal.

Greenspan also claimed that a failure (bankruptcy without bailout) could have had severe repercussions on markets and:

> ... substantial damage could have been inflicted on many market participants, including some not directly involved with the firm, and could have potentially impaired the economies of many nations, including our own.

Dowd opposed this claim on the ground that a private group made an offer to LTCM that, had the Fed not been waiting in the wings, must have been accepted by LTCM as better than nothing. Thus the issue was not one of failure of LTCM vs. bailout arranged by the Fed; the issue was actually bailout independent of the Fed vs. bailout as arranged by the Fed. Dowd admits that the terms of the Buffet team were harsh, but private business is often harsh and in that sense, the Fed should have kept out of it. It is hard to disagree with this viewpoint.

Would the damage to the economy (due to LTCM failing) have been as severe as Greenspan said? Dowd doubted that it would for the following reasons:

- The amount of money in the currency and bond markets was far more than that controlled by LTCM. In the words of Dowd: "The markets might have sneezed, and perhaps even caught a cold, but they would hardly have caught pneumonia."

- There would undoubtedly be a buyer for LTCM at some price. It never was going to go under.

- Even if LTCM did go into bankruptcy, it would only affect derivatives markets, and there is no evidence that it would have caused a global liquidity crisis.

- Even after a major shock, history shows that trading resumes not long thereafter.

- Most firms in the LTCM types of markets have a variety of protection mechanisms that would have reduced their exposure below what they might otherwise appear to be.

Dowd believed that:

> The Federal Reserve's nightmare scenario—a mass unwinding of positions with widespread freezing of markets—is thus farfetched, even in the fragile market conditions of the time.

A point not raised by Dowd is this. If the LTCM fund had reached the proportion where its failure could have a nightmare impact on the economy, and funds like the LTCM are subject to risk, why wasn't the Fed in there regulating the LTCM in the first place?

In summary, the only effect of the actions of the Fed was to find better terms for the managers and stockholders of LTCM – a role manifestly unsuited for the federal government. The grounds for this intervention appear to be baseless, and probably represent cronyism between the management of the Fed and management of LTCM.

ALBANIA'S PONZI SCHEMES

Bezemer provided a detailed analysis of the 1997 collapse of the Albanian economy caused by the collapse of economy-wide Ponzi schemes.[101] This contrasted sharply with its successful transition to a post-socialist transition country in the years 1992–1996. In that period, "inflation was contained, GDP increased, and unemployment decreased considerably." Bezemer referred to this stark contrast as the "Albanian Paradox:"

> Albania, the smallest and least developed of the Eastern European transition countries, is located by the Adriatic Sea, bordering on Greece to the south,

Macedonia to the East and rump Yugoslavia to the north. It has a 3.2 million population, 56% of which is employed in agriculture.

Bezemer analyzed the causes of the mania. Only a very brief summary is given here. The schemes prospered through a combination of "restrictive monetary policies, large capital inflows, and financial market policies that were very strict for official banks but extremely lenient for informal financial intermediaries." There were few official banks because of the strict requirements for the founding of banks in Albania. These banks operated under significant constraints. Monetary and budgetary policies limited the official supply of money to the economy, so that official savings interest rates were less than the rate of inflation:

> In the absence of well-developed stock and real estate markets, this induced people and firms to revert to transactions on informal markets. The demand for financial intermediaries was especially large because of a considerable cash flow to the population in the transition years.... These circumstances drove first business people, later many more non-entrepreneur citizens to the informal markets, where Ponzi-like firms and foundations had started to operate.... The striking contrast is that, while the authorities imposed very strict conditions on growth of the official financial sector, there were no regulatory impediments for those 'banking' with Ponzi methods.

The Ponzi schemes offered savers high interest rates. Toward the end, monthly interest rates on savings in informal markets reached as high as 50%:

> Ponzi firms employed effective marketing and advertising strategies with full use of the state television, extracting much of Albanian household money from under the proverbial mattress.... The absence of warnings from the government, the frequent appearance of pyramid managers and government officials side by side at public meetings and on television, and the association of pyramid managers with the Democratic Party lent state credibility to the schemes. Not only were much of the population's savings and (domestic and external) income 'invested', but many people took loans and mortgages on their houses or land in the expectation of quick gains. As a result, ... by early 1997, the total value of received deposits reached US $1.2 billion, or 50% of GDP.[102]

These gains were so large compared to wages that many people just stopped working:

The end to large-scale Ponzi operations came in January-February 1997. In the last quarter of 1996, interest rates had risen from 30 to 50% monthly, with some foundations offering 100% monthly. . . . In February they collapsed, swallowing a large share of the population's savings. Social unrest spread rapidly as masses of demonstrating Albanians demanded compensation from the government and strong suspicions existed about its involvement in the schemes. These protests, six weeks of looting, the plundering of army arms depots and the emergence of irregular, armed bands caused the government to lose control over the larger part of Albanian territory.

Bezemer's analysis of Albania was thorough and credible. But he left out one factor: greed. The longevity of a Ponzi scheme depends upon the ratio of the rate of payout to all participants vs. the rate of intake of new funds from new investors. The Albanian Ponzi lasted only a few months because this ratio approached 0.3 to 0.5 toward the end. Bernard Madoff's Ponzi lasted for years because he held payments to about 1% per month, and most investors reinvested their (virtual) dividends.

CORPORATE AND ACCOUNTING SCANDALS

For the past several decades, the major accounting firms – long known as the "Big Eight" and later consolidated into the "Big Five"[103] "have been embroiled in a series of scandals involving their failure to detect and disclose financial irregularities at companies they audited."[104] In some cases, it seems likely that they not only "failed to discover or disclose," but actively participated in fraud.

During the 1970s accountancy firms were criticized for failing to alert shareholders to the problems that led to the collapse of the Penn Central Railroad and for not reporting about the widespread payment of bribes by US-based multinationals to secure foreign business:

In the 1980s Peat Marwick gave Penn Square Bank a clean bill of health just before it collapsed under the weight of bad energy loans. Various accounting firms found themselves being sued by the federal government for their role in auditing the books of crooked savings and loan associations. This led to a series of settlements, the largest of which was the agreement by Ernst & Young in 1992 to pay a record $400 million in connection with about a dozen failed S&Ls. The following year Arthur Andersen agreed to pay the feds $82 million to settle charges in connection with the collapse of Charles Keating's Lincoln Savings and Loan Association.[105]

According to tradition, an accounting firm must maintain a distance between itself and the client corporation, in order that it may act independently to adhere to accepted accounting principles without being swayed into malfeasance by an undue loyalty to corporate management manipulations. However, this constraint limited the revenues of the big accounting firms. In order to increase their revenues, the large accounting firms expanded their business from mundane auditing into management consulting, which for many of the firms became a multi-billion-dollar business:

> The problem was that consulting put the firms in the role of quasi-insiders and flew in the face of the accounting profession's independence rules. For decades the SEC and Congress periodically raised concerns about the dangers of the industry's increasing involvement in consulting. The most serious reform effort was mounted in 2000 by SEC Chairman Arthur Leavitt, but the industry used its lobbying muscle to defeat Leavitt and later breathed a sigh of relief when Harvey Pitt, a corporate lawyer who had represented the industry, was named by President Bush to head the Commission: [106]

Mattera described the evolution of the new attitude of major accounting firms with examples of failure to disclose financial problems of Penn Central prior to its collapse in the 1970s, and various accounting firms were sued by the federal government for their role in auditing the books of crooked savings and loan associations in the 1980s. A number of accounting firms made settlements with the government, the largest of which was by Ernst and Young, who paid a record $400 million in 1992, in connection with about a dozen failed S&Ls. The accounting firms also had to respond to lawsuits brought by disgruntled shareholders of companies whose problems were not revealed by auditors. From 1980 to 1985, Arthur Andersen, paid out over $137 million to plaintiffs.[107] It has been reported that in the 1990s, the chicanery of accounting firms in collusion with corporations in the dot.com bubble reached unprecedented proportions.

As Forbes Magazine said: "With the avalanche of corporate accounting scandals that have rocked the markets recently, it's getting hard to keep track of them all."

Table 2.4 presents a brief glimpse at a small fraction of the corporate accounting scandals of the 1990s. This is just the tip of the iceberg.

It is difficult to perceive a pattern in the punishments meted out to financial culprits, and they do not seem to be proportional to the crime. While some financial criminals were sentenced to substantial terms (Adelphia, Enron, Rite Aid, DBL, Tyco, Worldcom) many others received unusually light sentences. In 2008, the ex-president of Peregrine Systems was sentenced to only three years

Table 2.4 Brief summary of a few of the many corporate scandals of the 1990s. (All data in this table were taken from news reports from legitimate news agencies, but no guarantees are made as to accuracy)

Company	Accountants	Issue
Adelphia Commun- ications	Deloitte & Touche	Five former Adelphia executives indicted on 24 counts of securities fraud, wire fraud, and bank fraud. Their actions, were described by a US Attorney as "one of the most elaborate and extensive corporate frauds in history." (See "Adelphia" in Chap. 2) John J. Rigas, founder, guilty verdict, 15 years Timothy Rigas, former CFO, guilty verdict, 20 years James R. Brown, former vice president, pleaded guilty, Awaiting sentencing
AOL-Time Warner	Ernst & Young	Several class-action lawsuits were brought against AOL-Time Warner in 2002 for allegations that it made material misrepresentations to the market, thereby artificially inflating the price of AOL Time Warner securities. The complaints alleged that they failed to disclose: (i) that the Merger was not generating the claimed synergies; (ii) that the Company was experiencing declining advertising revenues; and (iii) that the Company had failed to properly write down more than $50 billion of goodwill. These suits were consolidated into one, and in the final settlement, AOL-Time Warner was ordered to pay $2.65 billion and Ernst and Young had to pay $100 million into a settlement fund.
Bristol-Myers Squibb	Pricewater- house Coopers	Bristol-Myers Squibb agreed to pay $150 million to settle SEC accusations that the company improperly inflated its sales and earnings in a series of accounting frauds. In addition, Bristol-Myers agreed to pay $300 million to settle a shareholder class-action lawsuit over similar claims. The SEC said in addition, it would still pursue an inquiry that could result in civil fraud charges against individuals. Nor does the settlement resolve a Justice Department criminal investigation into the same accounting practices that could bring criminal charges against Bristol-Myers or its employees. The SEC assistant regional director who led the investigation said the substantial penalty was appropriate for the severity of the company's missteps: "'This is extremely egregious accounting fraud.... There

Table 2.4 (continued)

Company	Accountants	Issue
		was a $1.5 billion revenue recognition problem, which puts it second only to WorldCom."' Bristol-Myers used several earnings management techniques to distort the company's true performance from early 2000 until the end of 2001. Regulators say that Bristol-Myers inflated its revenues by more than $1 billion, going back at least to 1991.
Cendant Corp.	Ernst & Young	Cendant agreed to pay $2.83 billion to settle a shareholder lawsuit accusing it of fraud. Cendant had issued false and misleading statements and allowed former company directors and officers to sell Cendant shares prior to the disclosures of the accounting problems.
		Ernst & Young LLP settled for nearly $300 million a lawsuit brought against it by Cendant Corp., according to a securities filing late last year by a former Cendant subsidiary. In its suit, Cendant alleged that Ernst negligently failed to detect a massive fraud during its audits of a unit of the company.
		Ernst had already, in 2000, paid out $335 million to Cendant shareholders as a result of the fraud. That is believed to be the largest-ever settlement by an auditor related to work for a single client.
CMS Energy	Arthur Anderson	CMS was charged with: (i) deceiving the investing public regarding its business, operations and management; (ii) offering securities to the investing public at artificially inflated prices which incorporated false and misleading statements; (iii) engaged in illegal insider trading; and (iv) overstating revenue by nearly $5.2B in 2000 and 2001 by using artificial "round trip" energy trades. CMS agreed to a $200 million settlement of shareholder suits.
Computer Associates Inter- national	KPMG	Computer Associates was charged with "falsely reporting to investors and regulators during numerous fiscal quarters. . ., improperly recognizing and falsely reporting hundreds of millions of dollars of revenues. . .," as well as "obstruction of justice and perjury." The former chief executive of Computer Associates International, Sanjay Kumar, was sentenced to 12 years in prison for orchestrating a $2.2 billion accounting fraud at the software company. He was also fined $8 million.

Table 2.4 (continued)

Company	Accountants	Issue
Dollar General Corp.	Ernst & Young	Dollar General overstated its earnings $200 million from 1998 to 2000. Dollar General paid $162 million to settle class action shareholder lawsuits for the accounting missteps.
Duke Energy	Deloitte & Touche	Duke Energy and some of its team members were prosecuted for various crimes, including conspiring to drive up electricity costs in western energy markets, and various counts of conspiracy and fraud. Several of those charged were acquitted, but Duke reached a settlement agreement in principle with the states of California, Washington and Oregon; federal regulators; California's three largest investor-owned utilities; and other parties to pay $200 million in restitution.
Dynegy	Arthur Anderson	Dynegy Inc., is a Houston-based producer of natural gas and electricity that tried to take over a failing Enron but instead went into its own tailspin. It has been the target of several federal probes into alleged sham trades aimed at artificially inflating revenue and volume. Dynegy's longtime chief executive, resigned under pressure.
		The University of California announced a $474 million settlement on behalf of Dynegy investors in the securities fraud class action case of which it was lead plaintiff. This settlement includes $468 million from the company; $5 million from Citigroup, a bank involved in the fraudulent transactions; and $1.05 million from Arthur Andersen, Dynegy's auditor in 2002.
		Jamie Olis, a former midlevel executive at Dynegy and two former associates at Dynegy were found guilty of devising a secret project to disguise a $300 million loan as cash flow. Their sentences however, were modest.
El Paso	Pricewater-house Coopers	The FERC approved El Paso's $1.55 billion settlement with the State of California over allegations that El Paso manipulated natural gas prices during the state's electricity crisis. El Paso will pay $551 million in cash up front, and the remainder with additional payments and rate reductions.
Enron	Arthur Anderson	Enron manipulated the utilities markets by (1) spending great sums of money to influence regional legislators to pass deregulation policies

Table 2.4 (continued)

Company	Accountants	Issue
		favorable to Enron, (2) buy up control of suppliers of utilities in these regions where deregulation was in force, (3) use their control of the regional utility supplies to force up prices paid by suppliers to end-users (i.e. the public) and thereby make huge profits at the public's expense. They were eventually convicted of various crimes and the company fell apart. See "Enron" in Chap. 2.
		Kenneth Lay, former chairman, guilty verdict, deceased
		Jeffrey Skilling, former CEO, guilty verdict, 24 years
		Andrew Fastow, former CFO, pleaded guilty, 6 years
		Lea Fastow, former treasurer, pleaded guilty, 1 year
		Michael Kopper, former managing director, pleaded guilty, 3 years
Global Crossing	Arthur Anderson	It was claimed that Global Crossing swapped network capacity with carriers to boost sales and falsify earnings while former Chairman Gary Winnick made over $700 million selling stock. After the company filed for Chap. 11, Winnick pledged $25 million to help the workers who had lost their savings in the collapse while workers actually lost more than ten times that amount, tens of thousands of jobs were lost, and stockholders lost billions. Former VP of finance and whistleblower Roy Olofson claims that Global Crossing falsified more than $1 billion in revenue through round-tripping and that its auditor, Arthur Andersen, permitted improper bookkeeping entries. Global Crossing paid $325 million to settle a class action suit filed by former stockholders.
Halliburton	Arthur Anderson	Halliburton has been accused of numerous scandalous operations, none of which seem to have been resolved judicially, possibly due to influence of VP Cheney.
IMClone Systems	KPMG	Former CEO Sam Waksal paid $800,000 to settle charges that he was involved in $5 million of stock sales based on insider information; he allegedly dispersed information to family and friends, including Martha Stewart, regarding the FDA's pending rejection of the company's new cancer drug, Erbitux. Waksal also pled guilty to six charges including securities fraud, bank

Table 2.4 (continued)

Company	Accountants	Issue
		fraud, conspiracy to obstruct justice, and perjury. In his bank fraud plea, he confessed to forging a lawyer's signature for a $44 million bank loan.
Kmart	Pricewaterhouse Coopers	KMART has been involved in so many scandals that it is difficult to keep track of them all. Two former VPs were indicted by the SEC and DOJ for securities fraud. These were considered to be only the first results of investigations into a "company (that) collapsed in January 2002 under a team of executives known inside Kmart as the Frat Boys, who misused corporate jets, drove luxury leased cars, and received lavish salaries while steering the company into the largest retail bankruptcy in history."
		The Securities and Exchange Commission filed charges against two former top Kmart executives for misleading investors about Kmart's financial condition in the months preceding the company's bankruptcy.
		A federal judge approved a settlement in a class-action lawsuit filed against a group of former Kmart Corp. executives that gives approximately 125,000 employees and retirees of Kmart $11.75 million to settle their 2002 claims against the company executives. The US Securities and Exchange Commission accused the two men of making "materially false and misleading" disclosures to shareholders before the retailer's 2002 bankruptcy filing.
Merck & Co	Arthur Anderson	Merck recorded $12.4 billion of income through pharmacy benefits that it subsidiary, Medco, never actually collected. (Chicago Tribune, July 9, 2003) Merck has agreed to pay $42.5 million to settle class action lawsuits claiming that Medco "pocketed billions of dollars in rebates from manufacturers and other fees that they said should have gone to thousands of health plans and millions of consumers." Lawyers claim that Medco has kept more than $4.1 billion since 1995.
Micro Strategy	Pricewaterhouse Coopers	Three Microstrategy executives, CEO Michael Saylor, COO Sanju Bansal, and CFO Mark Lynch, paid $10 million in an SEC settlement alleging accounting fraud in December 2000. Each of the men was also fined $350,000, the largest fines in SEC history for cases not

Table 2.4 (continued)

Company	Accountants	Issue
		involving insider trading. Microstrategy repeatedly delayed or pre-booked deals to inflate company earnings. They converted a $40.3 million loss in 1999 to a $12.6 million profit using fake accounting. Microstrategy agreed to pay $97.0 million in a class action suit in Oct. 2000. The stock had lost $15.9 billion in value between March and the time the suit was settled.
Peregrine Systems	Arthur Anderson	Peregrine reported that it had misstated its income by $1.5 billion, but later revealed it had misstated by $4.1 billion. The company used several deceptive accounting techniques, including fictitious sales, mis-booked transactions, and deferment of revenue to invent two-thirds of its income. This information was not released until the week after four members of the board had already quit, including Chairman John Moores who made more than $611 million in stock sales since the company's initial public offering in 1997. Trials were still proceeding in 2007 but at least eight individuals pleaded guilty. However, the ex-president was sentenced in 2008 to only three years probation for lying about what he knew of the fraud that destroyed the software company.
PNC Financial Services Group	Ernst & Young	The settlement with the Justice Department required PNC to pay $115 million in penalties connected with securities fraud charges in July 2002. The charges related to PNC's effort to remove from its books $762 million in bad corporate loans and investments in 2001, which inflated annual earnings by $155 million (52%). The scandal also led to the departure of several key executives, including PNC's former Vice Chairman Walter Gregg Jr. and Chief Financial Officer Robert Haunschild.
Qwest Communications	Arthur Anderson	Qwest wrongly booked $1.1 billion in revenues between 1999 and 2001. Qwest used "round-trip" trades to inflate its revenues, which is the swapping of equal amounts of fiber-optic capacity with other companies only to inflate revenues. Eight executives have been charged by the SEC with involvement in booking $144 million in revenue early to meet earnings projections, and four of those executives have been indicted by the Justice Department for

Table 2.4 (continued)

Company	Accountants	Issue
		falsifying $33 million in revenue in 2001 as well as falsifying records for its auditor, Arthur Andersen. Fortune magazine named Qwest the greediest corporation in America because former CEO Joseph Nacchio and former Chairman Philip Anschutz made $2.2 billion in company stock sales just before the stock value fell. QWest agreed to pay $400 million to settle a class action suit, but that suit remained unsettled in early 2008. After telling former Qwest chief executive Joe Nacchio that he committed "crimes of overarching greed," a federal judge ordered him to serve six years in prison, pay $19 million in fines and forfeit $52 million in ill-gotten gains for illegally selling company stock.
Rite Aid	Deloitte & Touche	Rite Aid restated $1.6 billion of earnings in June 2002, the largest corporate restatement in history at the time. The former CEO, Martin Grass, and two aides were indicted by a federal grand jury on 37 counts of fraud, conspiracy, and lying to shareholders. The SEC has accused the three men of masterminding schemes to overstate income by cheating vendors, falsifying documents, and ordering executives to inflate numbers. These actions led to a 5,533% profit overstatement in the second quarter of 1999. Former Rite Aid CEO Martin Grass, drew an eight-year sentence for accounting fraud. Former Rite Aid chief counsel Franklin Brown is serving a 10-year term for his role in an accounting scandal.
Tyco	Pricewater-house Coopers	Former Tyco executives CEO Dennis Kozlowski and CFO Mark Swartz allegedly stole $600 million from their company by bribing their board to keep secret $170 million of unauthorized bonuses and loans and $430 million of stock sold at inflated value. Kozlowski used company funds to buy a $30 million apartment, a $21 million yacht, as well as a $17,100 traveling toilet box and a $15,000 dog umbrella stand. Kozlowski has also been indicted on a 14-count charge of $1 million tax evasion. Tyco has agreed to pay about $3 billion to settle shareholder claims from one of the largest corporate fraud cases ever. • Dennis Kozlowski, former CEO, guilty verdict, 8 to 25 years

Table 2.4 (continued)

Company	Accountants	Issue
		• Mark Swartz, former CFO, guilty verdict, 8 to 25 years
Waste Manage ment	Arthur Anderson	Former Waste Management CEO Dean Buntrock and five other executives allegedly made $29 million dollars from annual bonuses and insider trading while stockholders lost $6 billion by perpetrating $1.7 billion of accounting fraud during the 1990s, according to the SEC. The Waste Management executives worked with Arthur Andersen to undertake "massive earnings management fraud" which included hiding debts, overestimating property values, and not writing off the cost of unsuccessful or abandoned landfill projects. Arthur Andersen paid $7 million to settle a civil suit with the SEC that alleged false and misleading audit reports of Waste Management from 1993 to 1996. In March 2002, former Waste Management Inc. CFO James Koenig and five other former executives were sued by the SEC for their roles in the then-largest accounting scandal in US history. Koenig was sentenced to pay $4.2 million total in disgorgement, prejudgment interest, and civil penalties, and prohibited him from serving as a director or officer of a public company. A jury found him liable for violating securities laws 60 times between 1992 and 1997 in a scheme that resulted in a $1.7 billion overstatement in the Houston-based trash collector's profits.
World Com	Arthur Anderson	Worldcom has restated $11 billion for costs that it had wrongly booked, which eventually led to a record setting $107 billion bankruptcy. Yet, the SEC settled with Worldcom without fining it. Instead Worldcom agreed not to break laws in the future, despite the claim that Worldcom has committed the largest case of corporate fraud in US history which has cost investors over $176 billion. Amongst other things, executives created a scheme to book short-term operating costs in small increments as long term capital costs so they could meet revenue goals. Worldcom also set a record for the largest single write-off in US history when it restated $79.8 billion to re-estimate the value of its assets. Sentencing: • Bernard Ebbers, former CEO, guilty verdict, 25 years

Table 2.4 (continued)

Company	Accountants	Issue
		• Scott Sullivan, former CFO, pleaded guilty, 5 years • David Myers, former controller, pleaded guilty, 1 year, 1 day • Buford Yates, former accounting director, pleaded guilty, 1 year, 1 day
Xerox	KPMG	In July 2002 Xerox restated its 5-year revenue by $6.4 billion and its income by 36%, or $1.41 billion. Xerox inflated its revenue by booking deals before they were signed, and by keeping those pre-booked deals on the records even if they fell through. KPMG, Xerox's auditor, was sued by the SEC. for its involvement in fraudulent accounting practices for Xerox. Xerox paid the SEC $10 million to settle allegations of accounting fraud in April 2002 for its initial restatement estimate of $3 billion. The SEC levied record fines against current and former KPMG auditors accused of helping Xerox Corp. overstate revenue by $3 billion. The two men who directly oversaw Xerox audits, Ronald Safran and Michael Conway agreed to pay $150,000 each to settle a lawsuit brought by the SEC. The previous record was $100,000 that former KPMG partner Joseph Boyle agreed to pay to resolve his role in the Xerox audit.

probation for lying about what he knew of the fraud that destroyed the software company. A website[108] lists several cases where the punishment seems incredibly mild compared to the crime, and faults the SEC for tolerating corporate crime. One case in point provided by the "Skeptical CPA" was Broadcom, which:

> agreed to pay $12 million to settle SEC charges it falsified its reported income by backdating stock-option grants over a five-year period. The company restated its financial results in January 2007 and reported more than $2 billion in additional compensation expenses. . . .

The amazing thing here is that Linda Thomsen, the SEC's enforcement-division director, was quoted as saying that "the significant penalty imposed on the company" was warranted by the "scope and magnitude of the fraud." The "Skeptical CPA" responded with sarcasm:

Is Thomsen joking? A $12 million penalty for a $2 billion restatement! Wow, that's .006 of the restatement. I'm sure this will deter future corporate miscreants.

A New York Times article "points to an ideological sea change on the Supreme Court" who in the past "viewed big business with skepticism — or even outright prejudice" but now is "more receptive to business concerns."

> In a case in 1964, the court ruled that aggrieved investors and consumers could file private lawsuits to enforce the securities laws, even in cases in which Congress hadn't explicitly created a right to sue. In the mid-1990s, however, Congress substantially cut back on these citizen suits, and the court today has shown little patience for them.
>
> This term, the Supreme Court has continued to cut back on consumer suits. In a ruling in January, the court refused to allow a shareholder suit against the suppliers to Charter Communications, one of the country's largest cable companies. The suppliers were alleged to have "aided and abetted" Charter's efforts to inflate its earnings, but the court held that Charter's investors had to show that they had relied on the deceptive acts committed by the suppliers before the suit could proceed. A week later, the court invoked the same principle when it refused to hear an appeal in a case related to Enron, in which investors are trying to recover $40 billion from Wall Street banks that they claim aided and abetted Enron's fraud. As a result, the shareholder suit against the banks may be dead.[109]

For a number of years, Lloyd's Bank of London ran an ad that said: "At Lloyd's Bank, we know whom we work for." However, it was never clear whether this implied that they worked for the clients, the shareholders or the management. Fortunately, we have no uncertainty as to whom the Supreme Court works for – "we hold these truths to be self-evident."

> It is abundantly clear that the Supreme Court, the Federal Reserve and the Department of the Treasury are run for the benefit of the rich, and ultimately, we have a Government of the rich, by the rich, and for the rich.

Perhaps the biggest, most publicized scandal involved a number of players in the 1980s with names like Levine, Boesky, Milken and others. The story is long and involved.[110] One key player was Dennis Levine, who is described by Stewart as a rather incompetent investment banker with a big ego and ambitions to match. Levine promoted and nurtured personal connections to several colleagues in other investment banking companies during the 1980s when corporate takeovers were rampant, and big profits could be made by buying stock in the company to be taken over prior to the takeover event. Using inside information, he bought stocks via secret overseas accounts in Swiss banks, and parlayed a rather meager starting position into considerable wealth. Not only did Levine acquire information from conversations, he actually used a connection to steal into a rival investment house and rifle through their files repeatedly. He also worked closely with Ivan Boesky, who ran an arbitrageur business, to take positions in stocks ripe for takeover in order to force the hands of the companies to bend to their demands. What is interesting about Levine is that the several investment banks that he worked for thought him to be technically incompetent, but he seemed to have an uncanny ability to identify takeover situations before they became public. As a result, he made a good deal of money for his investment banks and he was highly valued even though he didn't seem to be very smart. But Levine claimed to have an almost prescient sense of the market and demanded high pay and exalted positions from the investment banks based on his performance – which was mainly based on illegally acquired inside information. Levine lived high on the hog with an extravagant lifestyle until he was caught. He arrogantly sneered at the SEC, which during the Reagan years, was under-funded and not encouraged to take any initiative. What ultimately did him in was that his operations became too conspicuous; the investigators found 28 clear instances of his insider trades prior to a takeover, and all his profits went into one overseas account.[111] After Levine was arrested, he told authorities of the relationship to Boesky, who in turn was arrested. Along with Milken, the triumvirate was guilty of insider trading, false public disclosures, tax fraud, and parking violations,[112] market manipulation and other assorted technical crimes. Both Levine and Boesky pleaded guilty; Boesky received a $100 million fine and both received prison sentences. It has been the experience of this writer that insider trading has always been present and even rampant on Wall Street. It is clear that stocks move early in anticipation of news releases, events and takeover bids. Some of this is legitimate anticipation by astute investment gurus, but a good deal of it is undoubtedly based on illegally obtained inside information that inevitably leaks out. Separating one from the other is difficult. If a given trader uses inside information only once, or rarely, it is almost impossible to catch him. It is only in the case of extreme arrogance where a trader carries out illegal trades repeatedly that a pattern emerges that allows him to be caught. That is what happened to Levine and his network of conspirators.

In the mid-1970s, standards for bonds were high, and most corporate bonds were rated as investment grade. The few bonds with low ratings that existed were mostly from companies that once flourished but had fallen on hard times ("fallen angels"). These bonds were discounted and thereby paid much higher current interest than investment grade bonds, because of the perceived risk inherent in what we now call "junk bonds." Michael Milken was influenced by studies of low-grade bonds that indicated that the risk inherent in these bonds was not so much greater than for investment grade bonds, and the additional interest could prove to be very rewarding. Milken went to work for Drexel, Burnham and Lambert (DBL) and by 1977 controlled a significant percentage of the then rather thin market in junk bonds. It is claimed that through careful research and selection, a diversified bond portfolio made up of junk bonds paid sufficiently high interest rates to more than make up for the slightly higher risk.

An innovation that Milken introduced was to actually underwrite new issues of junk bonds by companies in poor financial condition. As we stated, prior to Milken, such companies would not have had the temerity to offer junk bonds as a new issue – junk bonds had previously been mainly remnants of "fallen angels." But it was necessary to create funds to carry out Milken's ultimate plan, which was to use *funny money* obtained from a *junko* company selling junk bonds to the public, as a means of taking over legitimate companies with real assets, worth much more than the junk bonds that financed the takeover. As companies with poor credit found they could raise capital without having to offer equity shares, the junk bond boom took off. DBL had created a market for first-issue junk bonds. In 1983, DBL underwrote its first $1 billion junk bond issue, for MCI Communications. DBL's share of the junk bond market peaked at about 75% in 1983 and 1984.

Milken, like Levine and Boesky, had contempt for rules and regulations, and engaged in activities that may be viewed as unethical or in some cases illegal. When Boesky confessed, he implicated Milken in several illegal transactions, including insider trading, stock manipulation, fraud and stock parking. This led to an SEC probe of DBL. The probe went on for two years, during which Milken and DBL parried and obfuscated the investigations. In 1989, Milken was indicted on 98 counts of racketeering and fraud. In an effort to end what he viewed as government harassment, Milken finally pled guilty to 6 felony counts, including illegal insider trading, and filing false tax returns. Milken received a 10-year sentence but only served 22 months in prison. He remains a very wealthy man despite paying about a billion dollars in fines.

THE SUB-PRIME REAL ESTATE BOOM 2001–2007

ORIGIN OF THE BOOM

During the heyday of the dot.com stock market bubble, from about 1997 to early 2000, there was a pervasive feeling (and an induced actuality) of wealth amongst those who invested in the stock market. This led investors to be willing to invest in more elaborate housing, and as a result, house prices rose each year from 1997 to 2001. The Case-Shiller composite housing index increased from 78 in January 1997 to 115 in January, 2001 – an increase of 47% in four years. However 1997 house prices were depressed (see Figure 2.11; also Table 2.5).

When the dot.com stock market bubble deflated in the spring of 2000, there was a short hiatus in the price rise of housing. Then the Federal Reserve rushed in to try to preserve the stock market bubble, or at least mitigate the depth of the ensuing debacle. It seems likely that the Federal Reserve System has a primary goal to prop up ballooning stock markets, and was alarmed at the steep drop in the stock indices, particularly the NASDAQ in 2000–2001. The Federal Reserve believed that it had to act. The federal funds rate, which had wavered between 5 and 6% from 1994 to March 2001, was successively reduced to 4% in May 2001, to 3% in September 2001, to 2% in November 2001, and as low as 1.25% and even 1.0% in 2002 and 2003. While

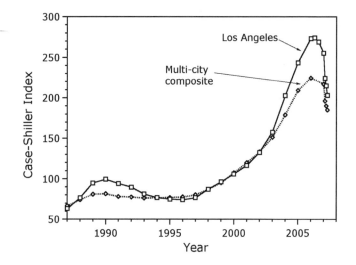

Figure 2.11 Case-Shiller real estate indices for Los Angeles, and for a composite of ten cities.

Table 2.5 New and existing single-family home sales and prices (Fannie Mae, 2007, updated 2008)

Year	New homes Sales 1000s of (units)	Median price	Average price	Existing homes Sales (1000s of units)	Median price	Average price
1989	650	$120,000	$148,800	3,010	$94,600	$119,100
1990	534	$122,900	$149,800	2,914	$97,300	$120,100
1991	509	$120,000	$147,200	2,886	$102,700	$130,000
1992	610	$121,500	$144,100	3,151	$105,500	$131,900
1993	666	$126,500	$147,700	3,427	$109,100	$134,700
1994	670	$130,000	$154,500	3,544	$113,500	$139,100
1995	667	$133,900	$158,700	3,519	$117,000	$141,500
1996	757	$140,000	$166,400	3,797	$122,600	$147,700
1997	804	$146,000	$176,200	3,964	$129,000	$156,700
1998	886	$152,500	$181,900	4,495	$136,000	$165,600
1999	880	$161,000	$195,600	4,649	$141,200	$175,200
2000	877	$169,000	$207,000	4,603	$147,300	$183,400
2001	908	$175,200	$213,200	4,735	$156,600	$192,900
2002	973	$187,600	$228,700	4,974	$167,600	$209,900
2003	1,086	$195,000	$246,300	5,446	$180,200	$225,000
2004	1,203	$221,000	$274,500	5,958	$195,200	$245,800
2005	1,283	$240,900	$297,000	6,180	$219,000	$267,400
2006	1,051	$246,500	$305,900	5,677	$221,900	$269,500
2007	775*	$247,300	$312,300	4,940*	$217,900	$266,200
May 08	590*			4,400*	$200,000*	$246,000*

* = roughly estimated from sources other than Fannie Mae.

the federal funds rate was later raised back to the 5% range by 2006, interest rates were extremely low from 2001 through 2004, and fairly low in 2005.

Almost every action of the Federal Reserve has had unintended consequences. It seems unlikely that the Fed desired to create a real estate bubble in America, but in case that was their aim, they succeeded handsomely. With the preponderance of existing mortgages at higher interest rates than the rates that prevailed for new mortgages in banks in the aftermath of interest rate reductions of 2001 to 2003, millions of households refinanced their mortgages during this period.[113] In doing this, many took advantage of the fact that their monthly payments would not increase if they raised the principal amount at the lower interest rate. Others raised the principal amount even more, using now fashionable adjustable rate mortgages (ARMs) with their "teaser" low initial interest rates.

While classical economists may think that lowering interest rates directly stimulates the economy, it seems likely that what actually happened in the early

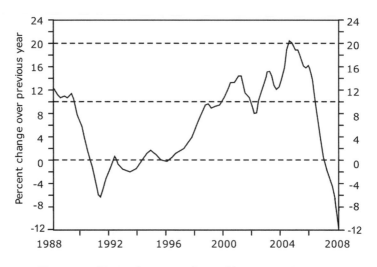

Figure 2.12 National average of annual house price increases.

2000s was that the lowered interest rates stimulated expansion of a real estate bubble, which allowed households to take cash out of their refinanced mortgages, and that was the stimulus for the economy from 2001 onward. Many households were willing to pay higher prices for residential housing because the monthly payments were manageable with the lower interest rates (although the piper would one day have to be paid on ARMs).

In this culture, where run-out cost is immaterial and only monthly payments are relevant, as interest rates dropped, more people were able to afford the monthly payments on more expensive houses. The demand for houses grew, and as a consequence, prices began to rise with gathering momentum.

EXPANSION OF THE BOOM

The national average of annual price increases is shown in Figure 2.11. For hot markets such as Los Angeles, the annual increases were even greater than the national average. The Case-Shiller index for Los Angeles increased by 18% in 2003, by 30% in 2004, by 20% in 2004, by 20% in 2005, and by 13% in 2006. What started out as a rise in house prices that mirrored the rise in stocks in the late 1990s, turned into a stampede to refinance and "trade up" in the 2000s. Somewhere around 2003–2004, the character of this rising market changed. Speculators moved in and began replacing those who buy homes to live in as their principal residences. Speculators bought homes with the intention of turning them over to a later-arriving speculator perhaps within a year or two. With low or even zero down payments, they had little to lose.

The quickest, easiest, most leveraged way to earn profits in this era was to buy a house and sell it a year or two later. In the speculative stage, the original reason for purchasing a residence is forgotten, and one invests only to turn over the investment to "a bigger fool." As the frenzy builds, speculators borrow to increase their leverage, and thus expand the bubble until it eventually pops. That is exactly what happened from 2003 to 2007. Many people bought several houses on little or no money down and sat back to await capital gains. Even more stretched their finances beyond the breaking point, knowing that they could not meet monthly payments for too long, with the expectation that a 10 to 20% per year price increase would bail them out.[114]

In a news report[115] it was claimed that in many cases, speculators lied on loan applications, saying they intended to live in the homes in order to obtain more favorable loan terms, and that roughly 20% of mortgage fraud involved "occupancy fraud," or borrowers falsely claiming they intended to live in a property. Borrowers who planned to live in a home could often purchase with no money down. These borrowers are much more likely to walk away from a mortgage and default if property values decrease, than homeowners who live in a house. Thus, many mortgages were more risky than agencies thought, and defaults piled up faster than expected when property values turned down in 2007. The article claimed that much of the occupancy fraud was concentrated in markets such as Florida, Nevada and Arizona, where prices were appreciating by double-digit percentages annually, and in Las Vegas, as many as 60% of the foreclosures in 2007 involved non-owner occupied homes.

More than one million US homes were in foreclosure as of early June 2008. This was by far, the highest rate ever recorded, and that number is predicted to continue to climb.[116] The Mortgage Bankers Association's first quarter 2008 report showed that a record 2.5% of all loans being serviced by its members were in foreclosure, which works out to about 1.1 million homes. The seasonally adjusted rate of homeowners behind on their mortgage payments also hit a record high. Nearly 3 million home loans, or 6.4%, had missed at least one payment, while about 737,000 were at least three months past due, but not yet in foreclosure.

All of this was done with the willing complicity of banks, agents, and appraisers. Mortgages were offered with no money down, with "teaser" low initial interest rates, sometimes with no principal paid off for the first few years, allowing deficits in payments to accrue to increases in principal owed, and with little or no checks on the financial status of the borrower, or the ability of the borrower to make payments. This aided and abetted those who saw purchasing a house as a way to make a quick profit. A frenzy of competition developed between banks to sell the greatest number of mortgages, however precarious the terms. Profits from new loan fees were the motivation. These loan fees were treated as current year income, boosting stock prices. As long as house prices kept rising, all excesses of judgment in granting mortgages would be forgiven. The Republican government's view was that *deregulation* of banks meant *no regulation*. The Fed pumped money into the system as fast as it was demanded.

As the traditional requirements for obtaining a mortgage were relaxed and in many cases, totally ignored, more and more "sub-prime" mortgages were issued that were extremely precarious, and depended on the expected 10–20% annual increase in asset value to bail out over-invested speculators. In addition to these sub-prime mortgages, Americans as a whole increased their debt via refinancing at higher levels, thus drawing out cash from their inflated homes (i.e. using their homes as ATM machines in which you only withdraw, but never deposit).

Figure 2.13 shows the percentage of zero down payment mortgages in California from 1998 to 2006. For first-time buyers, the figure reached 40% in 2006.

Prior to the 1980s, the typical householder would take out a mortgage with his local bank and if a temporary problem developed in making payments, he could go down to his local bank branch and work out a solution with them on a person-to-person basis. Since the 1990s, the more common thing has been that the local

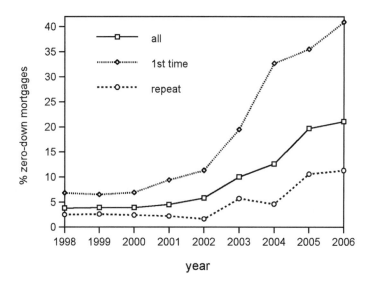

Figure 2.13 Percentage of zero down payment mortgages in California.[117]

bank branch will transmit your mortgage to the bank's central office, where they package up your mortgage with many others, and sell them to a much larger bank or mortgage investor. When you want to discuss this mortgage with the new owner of the mortgage, you can call them on the telephone and talk to a computer via a multitude of successive "press one" and "press two" actions, but it is difficult to gain access to talk to a human being, and it is impossible to talk to a local person – in person. As time went by in the 1990s and particularly in the 2000s, these packages of mortgages were converted into investment securities for trading, since they had some of the vestiges of bonds, generating interest income, and they seemed to be safe because they were backed by collateral in the form of housing. As the refinance mania expanded in the 2000s, a huge number of mortgages were created and most of these ended up in these packaged pseudo-bonds. The quality of these securities in many cases was not very good for a number of reasons, including (1) the increasing number of precarious sub-prime mortgages that were involved, (2) the susceptibility of payments on many mortgages to any downturn in house prices, and (3) the difficulty in dealing with homeowners scattered across the country who became delinquent in payments or who defaulted.

It is noteworthy that an IMF study[118] concluded that housing booms and busts were synchronized with stock booms and busts; however their data are suspect and when one compares Figures 2.4 and 2.13 for example, such correlation does not exist.

RESIDENCES AS ATMS

As house prices inflated, households withdrew cash from their houses in three ways: (1) by selling their houses at inflated prices, (2) by refinancing their mortgages at higher levels of principal, and (3) by acquiring "line of credit" loans in addition to their mortgage(s) using their houses as collateral. The cash that they took out of refinancing was plowed back into the economy, typically in the form of home improvements, personal consumption expenditures (PCE) – such as spending on vehicles, other consumer goods, vacations, education, and medical services – and also to pay off other forms of debt that had non-tax deductible interest. This provided a significant stimulus to the economy. The data on just how much of this took place are somewhat contradictory, but it is clear that the economy was greatly enhanced by this influx of cash. A 2007 Federal Reserve Report[119] provided the data in Table 2.6.

A website[120] reporting on a Greenspan release, provided slightly different (but similar) data:

Table 2.6 Cash proceeds from sale, refinance, and home equity loans on housing

Year	Cash proceeds from sale of existing homes($ billion)	PCE from sale of existing homes($ billion)	Free cash from refinancing($ billion)	Level of home equity loans($ billion)	Total cash proceeds per year($ billion)
1991	223.1	17.6	17.8	222.0	240.9
1992	175.4	13.8	25.3	217.1	200.7
1993	155.5	12.1	26.4	210.4	181.9
1994	173.8	13.5	17.4	221.8	191.2
1995	141.1	10.9	12.8	237.5	156.8
1996	206.0	15.9	21.8	262.6	231.1
1997	190.8	14.7	25.9	297.0	225.2
1998	249.9	19.2	46.9	309.9	262.8
1999	347.3	26.6	45.6	334.3	371.7
2000	389.0	29.3	32.6	407.4	462.1
2001	411.6	30.4	105.9	445.1	449.3
2002	488.2	35.3	140.2	501.1	544.2
2003	647.2	46.5	173.4	593.4	739.5
2004	701.5	48.7	146.2	778.4	886.5
2005	914.5	63.2	197.9	913.7	1049.8

Mr. Greenspan's new data show that borrowing against home values added a stunning $600 billion to consumers' spending power last year (2004), equivalent to 7% of personal disposable income – compared with 3% in 2000 and 1% in 1994.

In a speech given in 2003,[121] Alan Greenspan said:

All in all, the amount of previously built-up equity extracted from owner-occupied homes last year, net of fees and taxes, totaled $700 billion by our calculations, or more than 10 percent of estimated equity at the beginning of the year. Home equity extraction for the economy as a whole is, of necessity, financed by debt. In fact, the $700 billion of equity extraction is similar to the increase in mortgage debt last year.

Two Federal Reserve Reports[122] provide similar data. The 2000 report said:

Turning to the effect of cash-out refinancing, we estimate that, in total, $55 billion of equity was liquefied through cash-out refinancing in 1998 and early

1999. Survey findings suggest that about $10 billion of the $55 billion raised was used to fund activities that are classified in the national income accounts as consumption expenditures, such as the purchase of vehicles or other durable consumer goods, vacations, and education and medical expenses. Approximately $18 billion was used for home improvements.

The 2002 Federal Reserve report said:

> Turning to the immediate increase in the cash resources of the refinancers who liquefied home equity in 2001 and the first half of 2002, the average amount of equity withdrawn by these households was $26,723. Multiplying this figure by 4.92 million (the weighted 4.6 percent of the sample that refinanced and liquefied equity over the period multiplied by an estimated 107 million households in the United States) yields an aggregate estimate of funds raised through cash-out refinancing of $131.6 billion.
>
> $20.7 billion of the liquefied equity was used to fund purchases that are classified in the national accounts as personal consumption expenditures (PCE). An estimated $46.3 billion was spent on home improvements. Refinancers also used an estimated $28.1 billion to pay down non-mortgage debt and $5.8 billion to pay off second mortgages. Of the remaining liquefied equity, most (an estimated $27.5 billion) was invested in financial assets, real estate, or businesses.

As funds were drawn out of residences, personal savings plummeted and actually went negative, as shown in Figure 2.14.

The average American's savings rate (as percentage of disposable income) averaged about 10% from 1950 to 1985. However, as the stock market bubble expanded in the 1980s and 1990s, the perceived need to save was reduced and the savings rate plummeted, dropping to about 5% in 1993, 2% between 2000 and 2004, and turning negative in 2005.[123]

By the end of 2006, household real estate assets totaled about $20.6 trillion dollars, and with mortgage liabilities totaling about $9.8 trillion, household net real estate wealth totaled a little less than $11 trillion. This $11 trillion of net wealth represented about a 50% increase from 2000.

Thus, about $5.5 trillion of wealth was created out of thin air in six years – merely by bidding up the price of housing.[125]

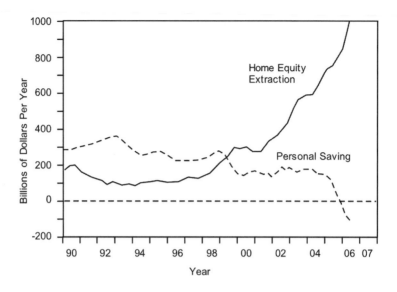

Figure 2.14 Annual extraction of equity from residence and personal savings 1990–2006[124].

The number of real estate licenses in California grew from 305,000 in early 2002 to 526,000 in early 2007.[126] By 2007, one out of every 50 adults in California had a real estate license!

Builders responded to this frenzy in the real estate market by building many more homes – often to be sold via shaky mortgages. Table 2.7 shows the number of

Table 2.7 Number of new houses built in the United States

Year	1000s of new housing units
2000	1,230
2001	1,270
2002	1,360
2003	1,500
2004	1,610
2005	1,720
2006	1,630
2007	1,380
2008	950 (est)

new housing units built in the United States starting in 2000. A great proportion of the increase was concentrated into "hot" areas.

The impact of the real estate bubble on the economy in the period 2002–2006 was significant. Whereas the contribution of the housing industry to job creation was typically about 10% from 1970 through 2000 (about 10% of new jobs), the contribution of the housing industry to job creation jumped up to 30% for the period 2000–2006.[127]

THE PUNCTURED BUBBLE

As Figures 2.11 and 2.12 shows, house prices topped out in late 2006 and early 2007, and accelerated downward in late 2007 and 2008. The drop from the 2007 peak was 24% in less than a year. This was the inevitable beginning of the end of the bubble. As house prices began to accelerate downward, we were treated to the usual reassuring comments by real estate organizations and banks. However, there was too much air in the balloon, and that air had to be released.

As we pointed out earlier in this book, a property of mathematics is that when a commodity's price increases by 100%, it needs only to drop by 50% to return to its original price. Thus, if a house starts off at say, $300,000 and increases by 100% to $600,000, it only needs to drop by 50% to return to $300,000. A house that goes up 200% in price only needs to drop by 67% to return to its starting price.

During the expansionary period of the bubble (2002–2007):

- A frenzy seems to have gripped the banking industry to market the greatest number of mortgages regardless of the generosity of the terms or the ability of the borrower to make future payments. Many of these mortgages were "sub-prime" to a greater or lesser degree.

- With the passage of time, as more and more of those with good credit had already purchased or refinanced, the banking and mortgage industry and builders turned to those with lesser credit ratings.

- While these mortgages were being marketed, profits to banks soared. Almost all attention was on short-term profits from up-front loan fees, while longer-term liabilities were ignored. Bank stocks rose.

- There was a huge expansion in mortgage-backed securities that were purchased by many leading banks and investment institutions.

- Some of these securities were insured by firms that insure bonds. It is not clear whether these insurance firms understood what they were insuring. Furthermore, in many cases, the insurers did not have sufficient assets to deal with a widespread debacle.

- Homebuilders responded to increased house prices by significantly increasing the number of homes built per year. Many of these new homes were concentrated in "hot" areas where the rate of home building was much greater than for the national average.

- "Flipping" became the modus operandi for buying houses. It is estimated that 20% of house purchases in California from 2002 to 2007 were by people who didn't live in the houses, and this percentage increased significantly from 2005 to 2007.

- After the bubble popped in late 2007:

- House prices began to fall. The rate of fall was highest in those regions where the bubble expanded the most. Florida, Nevada, Texas and California were epicenters for price reductions and gluts of unsold houses.

- The interest rates on adjustable rate mortgages with their "teaser" introductory rates began to increase at an accelerating pace.

- Marginal borrowers, depending on future double-digit price increases to cover their fundamental inability to meet payments, found themselves less able to meet commitments. The number of borrowers falling behind in payments soared, as did the foreclosure rate.

- The precarious nature of sub-prime mortgages and the lack of underpinning of mortgage-backed securities became widely known.

- The value of mortgage-backed securities plummeted, but it was difficult to appraise the true value of mortgage-backed securities, and trading in these securities was halted.

- The ability of insurance firms to bail out deficient mortgage-backed securities was questioned and became a major issue.

- As house prices decreased, buyers fled the market, leaving unsold existing and new home inventories at very high levels, which in turn, put a further drag on the housing market. Foreclosures added to the glut.

- One by one, major banks and securities firms reluctantly disclosed multi-billion dollar losses from holding mortgage-backed securities. Liquidity of the banking system came under question.

- The stock market, particularly the financial sector, after hitting highs in September 2007, dropped precipitously (but probably not nearly enough).

- Simultaneous puncture of real estate and stock bubbles boded poorly for the financial future.

It is noteworthy that a prescient prediction appeared on the Internet in February 2007.[128] Mr. Yones said:

> On January 31st, 2007, the president of the United States gave his speech on 'State of the Economy' citing strong economic growth, record Dow Jones performance and low unemployment rate. This report finds a different picture than the one announced. A deeper look into the economy reveals that the painted rosy picture is based on selective facts instead of a neutral assessment of all relevant numbers and economic trends. It is true that the US economy grew at 3.5 percent rate in 4th quarter of 2006, but the economic real growth is much less than advertised. Since 2001, economic growth has been largely fueled by rapid increases in asset prices (housing bubble) and expanding consumer debt rather than development projects, which results in non-sustainable and unhealthy (debt-driven) growth.

Yones went on to say:

> Any economy that is built on uncontrolled debt will eventually crash. . . . Many Americans refinanced their homes during the real-estate boom to pay for living expenses. With the expected housing bubble bust (declining housing values), Americans could lose a significant part of their savings.

A Wall Street Journal article[129] emphasized the willingness of borrowers to walk away from mortgage debt, contributing to extraordinary high levels of early default on loans issued during the 18 months before the mortgage bubble burst.

As the article pointed out, "a decade ago, most people started off with enough equity in their homes to make foreclosure irrational from a financial standpoint." With a 20% down payment on a properly appraised house, prices would have to fall incredibly before a homeowner lost all his money and had any incentive to walk away. Furthermore, as time went by, equity in the house increased, making fore-closure remote except in the case of an unexpected severe personal financial crisis.

However, the recent spate of foreclosures does not necessarily indicate that homebuyers are stupid. For a homebuyer who obtained interest-only mortgages for 5% down (or 0% down) in the expectation (or at least the hope) that a rise in house prices would bail them out, when house prices turned down, they had to consider walking away because they had more to lose than to gain by remaining with the mortgage. As the Wall Street Journal observed:

> Borrowers acted rationally in response to market forces and incentives during the bubble: Buy a house because 'prices always go up; you can't lose'. Many are acting rationally now: Mail the keys back and un-borrow the money, because prices are sinking fast while the debt isn't. When the house was purchased not as a first home but as a rental investment, the decision is even easier.

It is evident that in those regions where house prices went up the most in the era 2000–2006, they would come crashing down the fastest in 2007–2008. Southern California provides a good example. Figure 2.11 shows that the Case-Shiller index for Southern California increased from 100 to about 270 at its peak in early 2007. This was one of the greatest percentage increases in America. The average price of houses sold in Southern California rose from $415,000 in 2005 to a peak of $505,000 in February 2007.[130] Since then, it dropped to $278,000 in December 2008 – a drop of 45% from the peak. The number of homes sold in January 2008 was 45% lower than a year prior. Low interest rates, falling prices and promises of government relief were not enough to slow the pace of Southern California's housing downturn. Nearly 1 out of 4 homes sold had been foreclosed, which put additional downward pressure on home values. The number of residences in the final state of foreclosure in Southern California zoomed up from 337 in 2005 to 33,689 in the 2nd quarter of 2008.[131] By September 2008, half the houses sold were in foreclosure.

Double-digit price drops were recorded from early 2007 to mid-2008 in most major markets.

While many explanations are offered on business websites, few seem to emphasize excessive speculation as the cause. Once again, irrational asset price increases were treated as normal and only asset price decreases were considered to be abnormal.

The impact of boom and bust in residential real estate on financial institutions was dramatic. Almost every investment bank or mortgage seller revealed multi-billion dollar losses. Initial estimates of losses were around $150 billion, but with the passage of time, Goldman-Sachs raised the estimate to $460 billion in March 2008. Bill Gross, manager of the world's biggest bond fund, estimated that falling US home prices will force financial firms to write down $1 trillion from their balance sheets.[132] John Paulson, founder of a major hedge fund, said global write-downs and losses from the credit crisis may reach $1.3 trillion, exceeding the International Monetary Fund's $945 billion estimate.[133] Many details of the debacle are provided on the Internet.[134] This site provides a running record of events relevant to the debacles of major financial institutions in 2007 and 2008 with several hundred detailed entries and links to reports. Only a few brief examples are mentioned here. Countrywide Financial, a major issuer of mortgages saw its stock drop from the 40 s to under 5 and faced default until the Bank of America agreed to buy them out. Bear-Stearns, a major investment bank saw its stock drop from nearly 200 to 10, and was bailed out by intervention by the Federal Reserve. Lehman Brothers went into bankruptcy. IndyMac, a major bank for mortgages collapsed and was taken over by the FDIC. Wachovia's stock dropped from the 50 s to as low as 8 and Washington Mutual's stock dropped from the 40 s to as low as 1.7. The stock prices of both Citigroup and Bank of America dropped from the 50 s to as low as 15. Downey Savings and Loan stock dropped from 70 to as low as 1.3. First Fed Financial dropped from 60 to as low as 4. The stock prices of the two giant government-sponsored mortgage providers: "Fannie Mae" and "Freddy Mac" dropped from 82 and 67 to single digits, respectively, and eventually were taken over by the Government. Merril Lynch was rescued by Bank of America. When homeowners could no longer use their homes as ATM machines, and could no longer depend on stock-based retirement plans for their futures, they stopped buying products in the marketplace. Automobile sales dropped more than 30%. Oddly enough, the situation had parallels with 1988 when Eliot Janeway said: "By 1988, affluent Americans were sharing an unfamiliar insecurity with indigent Americans. The affluent were bracing for an onset of hard times, while the indigent were digging in to lower their standards of subsistence and to raise their tolerance of anxiety."

MORTGAGE-BACKED SECURITIES

In the 20th century, defaults on mortgages were rare, mainly because banks required a significant down payment, and house prices were relatively stable:

Federally chartered financial institutions such as Fannie Mae had been selling mortgage-backed securities to investors for decades. Those securities gave buyers higher yields than they could get on US Treasuries but also proved to be relatively secure investments. Even if one or two homeowners defaulted on their mortgages they represented a small fraction of the total mortgages packaged in such securities. And with ever-rising real estate values the chances were good that the full value of a defaulted loan could be recovered.[135]

It was then proposed to "offer investors an even higher rate of return by packaging the mortgages of less creditworthy homebuyers who could not qualify for a standard mortgage but were willing to pay a higher interest rate to become homeowners." Thus the biggest financial institutions in the United States began to purchase, package, and sell *Structured Investment Vehicles* (SIVs).

These investment vehicles were offered at a time when real estate prices had risen very sharply and there was a serious risk of a bubble-bursting decline in home prices. Such instruments added fuel to the fire of seemingly ever-rising home valuations. In addition, as Lilly said, *diligence* was ignored in evaluating mortgage applications, and

> ... loan originators had little or no stake in whether applicants had the financial capacity to repay the mortgages for which they were applying. The more applications they approved the more money they were able to make (via loan fees).

After the Federal Reserve cut interest rates in a desperate attempt to stem the stock market collapse of 2000–2001:

> ... millions of homeowners refinanced and millions more found that low rates permitted them to enter the housing market for the first time. There was a huge expansion in the mortgage origination business but by 2004 the flow of new applications began to subside. Mortgage originators needed to find new markets if they were to continue to collect the fees that kept them in business. The only real option was to expand the market by turning to the so-called sub-prime borrower — people whose financial history and current economic situation would have previously not permitted them to get a mortgage.[136]

Lilly asserted that even though it was known that:

> ... the mortgage originators had no incentive to insure the creditworthiness of the new borrowers, it was less well understood that at least some of the Wall

Street firms that packaged the sub-prime loans and sold them to unwitting investors were in much the same position.

Lilly quoted Allan Sloan, of Fortune Magazine, who reviewed the evolution of a package of 8,274 second-mortgage loans in the spring of 2006 by Goldman, Sachs & Co. This represented about 1% of the total value of the 916 residential mortgage-backed issues sold in 2006. However, it was, as Lilly said:

> ... clearly one of the worst. The average equity held by the homeowners[137] in this mortgage package was less than 1%. About 58 percent of the loans had no documentation or minimal documentation. No one knows whether these borrowers actually occupied the residences they were using as collateral, whether they were employed or whether they had any of the assets they told the loan originators they possessed.
>
> Goldman Sachs purchased these mortgages and packaged them into something called Goldman Sachs Alternative Mortgage Product (GSAMP) Trust 2006-S3. They then divided the $496 million in mortgages into 13 separate securities; three containing the "best quality" but lowest yield loans; seven containing the intermediate loans and the three containing the lowest quality but highest yielding loans. They then sold at least 12 and perhaps all of the 13 securities, which Sloan refers to as 'financial toxic waste'. The default rate on these loans has been so high that all three of the lowest quality securities are totally worthless, four of the seven mid-level securities are worthless and one other is deteriorating rapidly. The ratings on the top-level securities have been reduced from AAA to BBB and as a result their value has declined markedly. To date, the losses to Goldman Sachs customers are probably in excess of $300 million. But the real bombshell in Sloan's story was not the shockingly poor quality of the products that were sold or the massive losses that were absorbed by hapless buyers. The real surprise is that Goldman Sachs not only absorbed none of the losses, but in fact profited handsomely from the demise of the securities that they were telling clients to invest in. How? Because another part of Goldman Sachs was heavily shorting these securities in their own portfolio at the same time they were recommending them for the portfolios of other institutions.

Roger Lowenstein[138] provided a detailed review of the role of the ratings agencies in assigning financial ratings to the securities backed by mortgages. Almost all of the sub-prime mortgages ended up being pooled into mortgage-backed securities. The mortgages were sold to a "dummy" investment vehicle that received all the payments

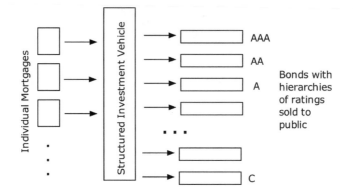

Figure 2.15 Mortgages packaged into a SIV that issues a hierarchy of bonds over a range of financial ratings.

made by mortgages that it held, as shown in Figure 2.15. The bonds with the highest ratings had first call on assets if enough mortgages defaulted that the SIV could not pay bondholders with income from mortgage payments; these paid the lowest interest. The bonds at the bottom of the ratings ladder would be the first to lose value in such an event (these paid the highest interest) and as more mortgages failed, the level of failing bonds would propagate upward. The ratings assigned to the bonds by the ratings agencies were extremely optimistic for several reasons:

(1) The rating agencies were making huge profits from the high volume of SIVs and were motivated to maintain the flow of SIVs, which depended on good ratings to be successful.

(2) If one rating agency did not provide sufficient ratings, the organizers of the SIV could "shop around" to seek a better rating from a competitor.

(3) The ratings agencies seem to have had their heads in the sand and were blissfully unaware of the level of speculation going on, the lack of veracity of data on borrowers, and the inadequate reviews of borrowers made by lending institutions. As a result, they used formulas and standards more appropriate to the pre-2000 era when house prices were more stable, and borrowers had to demonstrate to lenders their ability to make payments.

One major piece of stupidity was that the agencies tended to look only at the size of the first mortgage, but most borrowers also took out a second mortgage, so in effect they had no equity in the house.

As Lowenstein pointed out, the end result was that Moody's rating service had to downgrade more than 5,000 SIVs in 2007, resulting in huge losses to investors.

Wall Street icon Henry Kaufman, warned that the sub-prime problem is only part of a far larger problem in which giant financial conglomerates contribute to opaqueness in our financial markets. Kaufman argued that the Federal Reserve and US Treasury Department have failed to keep pace "with a series of fundamental structural changes that have transformed markets in recent decades."

As early 2008 wore on, the number of beleaguered institutions, and the extent of their losses continued to grow. When Bear Stearns became insolvent, the Federal Reserve moved in to arrange a rescue. After all, the Fed is nothing if not loyal to its financial cronies.

GOVERNMENT RESPONSE TO THE PUNCTURED BUBBLE

The response of the Government was fairly predictable. The Democrats expressed strong concern for the homeowners who were losing their homes due to foreclosure. In this respect, they showed incredible naiveté because they ignored (or were not aware of) the degree of speculation that had occurred, or perhaps they craftily utilized these speculators as political instruments, describing them as poor homeowners.

The Republicans were concerned for the banks and home construction companies that were suffering large losses; in addition they did not want the economy to slip into a recession in a major election year. The following actions were taken:

- A good deal of governmental jawboning was addressed to mortgage companies and banks asking them not to impose contractual increases in interest rates on ARMs (thus increasing losses at these companies).

- Press releases claimed that new funds were being made available by the Fed to banks to maintain liquidity.

- The Federal Reserve made a series of dramatic interest rate cuts, in each case immediately after each precipitous one-day drop in the stock market that was reacting to losses inflicted by the real estate market.

- The Congress, jointly by Democrats and Republicans alike, with support of President Bush, provided a "relief package" whereby most households would receive a payment from the government of up to $1,200 and would be encouraged to spend, spend, spend that money. This would total up to perhaps $160 billion (or more). It was not clear where these funds would come from, but it seems likely that it would be borrowed from foreigners, mainly the Chinese. However the Chinese would get a return on these funds because almost all the products for sale in America are made in China.

- Millions of American professional people, having experienced a stock market bubble since 1982, and being assured that stocks were the best medium to invest their 401(k) retirement funds, began to be concerned about their future retirement prospects.

A news report[139] said:

> Hillary and Bush agree: Government should bail out homeowners. Democratic presidential candidate Hillary Clinton called for a 90-day moratorium on foreclosures for homeowners who default on sub-prime mortgages. The New York senator, is also seeking a five-year freeze on the monthly rate for sub-prime adjustable mortgages. . . .

From early 2008 to July 2008, the Congress debated what to do about the sub-prime mortgage crisis. Finally, a mortgage relief bill was passed in July 2008 that was reluctantly signed by President Bush. This bill was a desperate (but inadequate) attempt to prevent the housing bubble from descending further. It provided government backing for "Fannie Mae" and "Freddie Mac," and funneled government money to housing speculators who faced foreclosure.

In an ironic twist, Arthur Greenspan, the architect and founding father of the sub-prime crisis, warned on July 31, 2008 that the real estate markets, already down significantly, had a lot further to drop. The stock markets responded by dropping precipitously in the next few minutes.

Despite Government efforts to prop up Fannie Mae and Freddie Mac, they finally collapsed in August 2008, and had to be taken over by the Government at a cost of several hundred billion dollars. Soon afterward, the Government took over AIG Group, the 2nd largest insurer in the United States.

INTERNATIONAL MORTGAGE DEBT

The great increase in mortgage debt of the 2000s was not limited to the United States. In September 2007 Morgan Stanley warned that the US mortgage crisis may precede a blowout of the entire European mortgage bubble. Belgium, Denmark, Greece, Great Britain, Sweden and Spain have undergone a very high growth in housing prices since 1997, which is even more unbalanced than the situation in the United States, when compared to population growth, income levels, and cost of money. The situation in Spain was identified as particularly critical.

Average house prices rose from 1996 to 2006 by 114% in Great Britain, by 133% in Spain, by 131% in Sweden, and by 90% in Belgium.[140]

A report by the International Monetary Fund[141] warned that home prices in many other industrial countries were even more overvalued than in the United States. The IMF attempted to assess how much house price increases could be justified in terms of economic fundamentals, and reached the conclusion that housing is even more overpriced in some countries other than the United States. The basis for the IMF study was a comparison of mortgage debt with the gross national product (GNP) for 17 countries. A steep rise in the ratio of mortgage debt to GNP has occurred in many countries over the past decade or two. The IMF then analyzed the economies of these countries and attempted to estimate the percentage of overpricing of housing in each country. (See Table 2.8).

There are a few countries where ratio of mortgage debt to GNP remains low, such as Austria (20%) and Finland (40%) and housing is even claimed to be under-priced in Austria.

However, it is far from clear how the IMF estimated the percentage overpricing in each country, and their results appear to be grossly understated for the United States. They claim that US housing is 11% overpriced, but as Figure 2.11 shows, the price of housing in the United States more than doubled from 2000 to 2007. Does that mean that housing was 50% under-priced in 2000? In some localities, such as Los Angles, housing increased by 170% from 2000 to 2007. Surely, housing in the United States is overpriced by far more than 11%?

The IMF also recommended that central banks should pay close attention to home prices and raise interest rates when prices are rising rapidly. It was stated that:

Table 2.8 IMF estimates of home-price overvaluation

Country	Ratio of mortgage debt to GNP in 2006	Percent home-price overvaluation
Denmark	100%	18%
Netherlands	98%	29%
Great Britain	80%	28%
Australia	80%	24%
United States	76%	11%

[This] conclusion is directly contrary to the established policy of most central banks, including the Federal Reserve, which ignores home prices when they are expanding. In the current credit crisis, which began with problems in the sub-prime mortgage market, the Fed has moved aggressively to lower interest rates.

But as we have amply demonstrated in this book, the Federal Reserve has a policy of promoting, supporting and sustaining bubbles, and the housing bubble of 2000–2007 is no exception. Evidently, the Fed believes that a doubling of house prices in 7 years is a good thing for America, and the Fed will do all it can to support this bubble.

THE HIDDEN TIME BOMB

While almost all attention to the sub-prime mortgage mess has been focused on structured investment vehicles and losses by major financial institutions, there remains a hidden time bomb for the economy that may take a few years to fully unfold. A large portion of the prosperity (and consumer spending) of the 2002–2007 period was based on huge borrowings by homeowners against their rising home values (as well as paper increases in stock prices), while average real wage increases have been modest. These increases in debt produce a burden on many households, and in a period of declining real estate values, consumer spending is likely to retract sharply. In short, homeowners will not feel as "rich" as they did in 2006. The economy is tanking in late 2008.

Finally, the years of living on borrowing and increasing paper assets have come home to roost in 2008. One of the old jokes on Wall St. is the investor who buys a penny stock and sees that it went up in the morning paper. He buys some more and notes that once again its price has increased in the next day's paper. This goes on for weeks; he keeps buying and price keeps going up. Finally, he calls his broker and tells him to sell the lot. The broker says "To whom?" Karl Marx identified the real problem with capitalism: Capitalism has the means of production but not the means of distribution.

As homebuilders amply demonstrated from 2002 to 2007, they have the means to build a gazillion homes. Out in the so-called "Inland Empire" 50 miles east of Los Angeles, they put up hundreds of thousands of tract homes per year. Outside Las Vegas, and in Florida, likewise. This country has the lumber, the copper, the iron, the supplies, and the labor to make so many homes it would make your head swim. Similarly, we have steel, and rubber, and plastic and metals and assembly plants that can turn out cars galore. And if they stopped making SUVs and made mainly Toyota Corollas, they would probably almost double their output. The problem is that the people who they would like to sell to, do not have the money to buy these homes and cars. So, we have no great problem with production — our problem is distribution. Until recently, we got around this problem by bidding up paper assets. Rising stock markets and low interest rates convinced Americans not to save because their futures were guaranteed by rising stocks and the income from savings was paltry. Instead, they spent their disposable income, and borrowed to buy even more. Rising house prices encouraged millions of Americans to use their homes as ATM machines, and by adding to their mortgage debt, they were able to generate cash that fueled an expanding economy. Many millions of others speculated in buying new houses with the intent of turning them over for a quick profit. As long as paper assets kept rising, all was well. Then the bubble popped.

The problem for the rich is how to get enough money into the hands of the people to buy the products that the rich produce, while remaining rich. My friend, Giulio Varsi, claims that the general approach that has been used is to provide welfare to the poor to give them cash to open up new markets for products, while maintaining low taxes on the rich to ensure their continuing wealth. The middle class bears the tax burden. As I have shown in this book, when you sum income tax plus social security, and take into account the proportion of income vs. capital gains, the total taxes on the first $100,000 of income are the highest of all income brackets.

The reason that we are unable to distribute houses and vehicles to all the people is that the money in America is concentrated in the hands of the rich. If the US really wants to distribute houses and cars to the wider populace, it is going to have to take the money away from the rich and give it to the people. That seems unlikely to occur.

JAPAN AND EAST ASIA

JAPAN 1970–2007

BACKGROUND

K&A provided a good historical background as to how Japan industrialized and opened up to foreigners in the late 19th century. Japan emulated the West in developing its railroads, its civil service, its banking system, its central bank, and its economy. The industrial economy in Japan developed around a limited number of feudal families such as Mitsui, Mitsubishi, Sumitomo, and Yamoto, each of which formed an industrial group. These industrial groups were outlawed in the late 1940s by General Douglas MacArthur. Nevertheless, after the groups were ostensibly split up, they continued to practice their inbred policies. As K&A said: "The Mitsui Steamship Company purchased its steel from the Mitsui Steel Company and its insurance from the Mitsui Insurance Company."

Japanese economic growth in the half-century after WWII was phenomenal. In the early years of this period, the quality of Japanese products was generally inferior. However, they improved greatly after the 1960s. K&A said:

> By the 1980s Japan was the second leading industrial power, more economically powerful than Germany. Toyota, Nissan, and Honda were leaders in the global automobile industry. Sony, Matsushita, and Sharp and a seemingly endless list of firms dominated the global electronics industry. Nikon and Canon 'owned' the world's photo-optics industry. Japanese-built computers were among the most powerful in the world.

With growing prosperity, real estate and stock prices began escalating rapidly in 1985, while there was also a rapid appreciation of the Japanese yen. Japan maintained low interest rate ceilings on both bank deposits and bank lending rates. The demand for loans from business firms at these low interest rates was much greater than the supply; loans were awarded on a government-directed preference basis.

The real rates of return on bank deposits and most other securities were negative. However, the real rates of return on real estate and stocks were positive and high. Hence more and more funds poured into the stock and real estate markets. This is similar to what happened in the US from 2002–2007.

The Bank of Japan reduced interest rates further after 1986, stimulating even greater boom conditions. However, prices of goods and services in Japan did not escalate excessively because of appreciation of the yen, which moved up from almost 240 to the dollar in 1985 to 130 in 1988. According to K&A, "deregulation of financial institutions was a major contributory factor to the asset price bubble in Japan in the 1980s and especially the second half of that decade." Japanese banks were engaged in a competition to acquire the most assets and the greatest number of loans. If this sounds familiar, think of the US in 2002–2007.

THE JAPANESE BOOM AND BUBBLE

Traditionally, Japanese firms had not been as concerned with bottom-line profitability as much as US firms: their priorities were to expand their product lines and provide lifetime employment for a growing number of employees. Market share was an important cultural measure, and many firms increased the amounts borrowed in efforts to improve product lines and increase their market share.

As the market value of Japanese stocks surged upward, investors resident in the United States and Western Europe bought more Japanese stocks. Foreign investors benefited from the combination of the increase in the price of the stocks and the increase in the foreign exchange value of the Japanese yen.[142]

The Nikkei stock index climbed from 6,000 in the early 1980s to 10,000 in 1984, and thence to almost 40,000 in late 1989. The price/earnings ratio jumped from under 25:1 earlier in the decade to over 60:1 in 1989. All of the financial values in Tokyo were sky-high toward the end of the 1980s. The market value of Japanese stocks was twice the market value of US stocks, even though Japanese GDP was less than half of US GDP. The comparison between Japanese and United States firms in terms of the ratios of the market value of stocks to profitability was even more skewed.

At the same time that stocks were advancing, property prices were increasing at the rate of 30% per year. According to K&A:

Real estate prices increased much more rapidly than rents, with the conse-quence that the rental rate of return declined significantly below the interest rate on the borrowed funds. Investors who had bought properties in the last several years of the 1980s had a negative cash flow - the rental income on their properties after the payment of the operating costs was below the interest payments due to the lenders - but because property prices were increasing so fast, they could raise cash to make the interest payments either by increasing the amounts borrowed against a property in earlier years or by selling.

This real estate boom was a predecessor to the sub-prime mortgage boom on the 2000s in the United States (and globally as well). In both cases, investments in unaffordable, highly priced real estate were made possible by annual double-digit increases in asset prices – that is, until the bubble popped.

K&A suggested that the bubble in Japanese real estate prices resulted from four factors: (1) tradition that land is a good investment, (2) the fact that real estate had been a winning investment for 30 years, (3) liberalization of constraints on banks to increase the proportion of loans for real estate, and (4) the rapid growth in the money supply in the second half of the 1980s as a result of the intervention of the

Bank of Japan to limit the appreciation of the yen in the foreign exchange market, which would have hurt exports.

As K&A explained, "firms involved in the real estate business accounted for a significant proportion of the market value of all of the firms listed on the Tokyo Stock Exchange." As real estate prices rose, Japanese banks (which owned large amounts of real estate and stocks) were able to increase their loans.

Japan appeared to have developed the financial equivalent of a *perpetual motion machine*. The increases in real estate prices led to increases in stock prices; the increases in both real estate prices and stock prices led to increases in bank capital.

As long as real estate prices continued to rise, the banks were solvent and all was well:

Industrial firms began to borrow to obtain the funds to buy real estate and shares in other firms because the rates of return were so much higher than the rates of return from [merely] producing automobiles and electronics and steel.

Thus, Japan went through a classic bubble in which the rise in asset prices dwarfed increased profits from ordinary enterprises, inducing more and more money to be invested into paper assets with the hope of further gain. During these boom years in the late 1980s, newly rich Japanese were almost giddy in flexing their financial muscles across the world. Japanese bidders bought French impressionist paintings at auctions, driving up art prices to new records. Expensive golf courses mushroomed. The Japanese competed to purchase the world's major real estate. According to K&A:

The Mitsui Real Estate Company paid $625 million for the Exxon building on Sixth Avenue in New York City against an asking price of $310 million because the company wanted to get into the Guinness Book of World Records. Other Japanese firms were also acquiring trophy properties and buildings in the United States. Mitsubishi Real Estate bought 50 percent of the Rockefeller Center, and a group related to Sumitomo Bank bought the Pebble Beach Golf Course in Northern California. Sony bought Columbia Records and then Columbia Pictures, and Matsushita, its dominant rival in the electronics industry, acquired MGM Universal.

The market value of Japanese real estate was twice the market value of US real estate, even though the land area in Japan is 5% of that in the United States and 80% of Japan is mountainous.

It was claimed by one estimate that the market value of the land under the Imperial Palace was greater than the market value of all of the real estate in California.

As K&A explained,

> The Japanese had all the money - and they were spending it to buy all kinds of assets both at home and abroad. The paradox was that the Japanese were spending as if they were very rich and yet there didn't seem to be that many rich Japanese; much of the spending was by Japanese corporations.

Smith[143] provided additional insights. He pointed out that as in the United States, Japan abandoned the traditional valuation of stocks in terms of dividend payments, in favor of valuation based on subjective beliefs in putative future earnings potential (see "The Valuation of Common Stocks" in Chap. 1).

Smith emphasized that in the early 1950s, roughly 70 percent of Japanese stocks had been held by individuals, but by 1989, 70 percent of listed shares in Japan were held by corporations, banks, and insurance companies, "who were unlikely to sell at any price." As Smith put it: "Japan, Inc., owned 70 percent of itself." The number of free shares on the market available for trading was limited.

Smith claimed that the Japanese stock market was fundamentally different from the United States and Western markets. The Japanese markets were institutionalized. Government intervention maintained high stock prices in the 1980s so that corporations had a source of cheap capital by selling stock at high prices. Smith therefore disputed the notion that "the Japanese stock market was an irrational bubble, inflated by investors acting on emotion rather than reason." He further argued that the effective number of shares was less than the nominal number due to extensive cross-holdings by institutions, and thus the earnings per share should have been increased proportionately, thus reducing the price/earnings ratios of stocks below the nominal values. In addition, unlike the United States where every penny increase in earnings seems to drive a stock wild, Japanese companies were under very little pressure to report high earnings, and often used accounting gimmicks to understate earnings so as to reduce taxes. Hence Smith argued that the concept of a herd-mentality, investor-driven bubble was not accurate for Japan. However, he did admit that because of the relatively small number of shares traded, it was possible for "unsavory market operators to manipulate prices" by cornering the market on shares. But merely because shares were held by institutions, we cannot conclude that the shares did not exist, and therefore the argument that the P/E ratio should have been lowered does not make sense to this writer.

According to Smith:

For most of the 1980s, the Japanese system seemed to be working. Stock prices rose at an annualized rate of 21 percent per year, despite the fact that operating profits per share for Tokyo exchange-listed stocks grew only 2.8 percent per year. (Net income for these firms rose at a 5 percent rate, but the heavy issuance of new shares reduced the increase in profits per share to only 2.8 percent.) It was thus the expansion of valuations (P/E ratios) that provided most of the gain in stock prices.

However, it is difficult to understand how this system was "working" when stock prices were appreciating ten times faster than earnings. How long could stocks continue to appreciate at 21% per year? As is usual in bubbles, there were many prognostications that things were different in this case, and we had traversed to a new paradigm. Smith pointed out that a warning sign was the fact that much of the reported earnings of companies in the late 1980s were due to capital gains from investments in shares, and not from operating profits.

A good deal of foreign money was invested in the thin Japanese stock market in the 1980s, but foreign money began to be withdrawn as the markets soared in the late 1980s.

COLLAPSE OF THE JAPANESE BUBBLE

The bubble in Japan reached its peak at the end of 1989. Banks developed 100-year, three-generation mortgages to deal with the high prices of real estate. However, the Bank of Japan was concerned that such high prices for homes had become problematic for the populace. Thus, a new Japanese central bank regulation limited the rate of growth of real estate loans.

Rental incomes in Japan were insufficient to cover interest payments on owners' mortgages, but owners managed by continually increasing borrowing based on continual increases in real estate prices. However, when the rate of growth of bank loans slowed, recent buyers of real estate developed a cash bind; they could no longer obtain the cash needed to pay the interest on their outstanding loans via new bank loans. Some of these owners therefore became distress sellers because of the high carrying costs. The combination of the sharp reduction in the rate of growth of credit for real estate and these distress sales caused real estate prices to decline.

Stock prices and real estate prices began their long decline at the beginning of 1990; stock prices declined by 30% in 1990 and an additional 30% in 1991. The stock price trend in Japan continued downward although there were a number of significant rallies. The Nikkei stock average dropped from a high of almost 40,000 at the end of 1989 to about 23,000 in 1991, then to about 19,000 in 1994, and trending down to about 15,000 by 2001. It bottomed out at around 10,000 in 2003, recovered to over 17,000 in 2005, and dropped to around 8,500 in 2008.

The decline in asset values made many Japanese financial institutions precarious. The banks became unwitting owners of thousands of French paintings. Many golf courses went bankrupt. Economic growth plummeted. The failures of firms meant that the banks took over title to the properties and sold them, putting further downward pressure on the price levels. So there was a downward spiral. Commercial and industrial enterprises went bankrupt at a steady rate of 1,000 per month. K&A called this: "debt, deflation, default, demography and deregulation."

As K&A described it, "the perpetual motion machine began to work in reverse." As property prices declined, bank capital declined and banks were now much more constrained in making loans. Since Japanese stocks were declining while US stocks were booming, investors sold Japanese stocks and bought US stocks. Bankruptcies increased, and the banks incurred large loan losses. "For the first time the banks began to ask: If we make this loan, what is the likelihood that we will be repaid?"

The parallel with the sub-prime fiasco in the United States of the 2000s is uncanny. From 2002 to 2007, US banks and mortgage companies seemed only intent on selling the greatest number of mortgages – regardless of the prospects for repayment, under the assumption that rising real estate prices would bail out all weak loans. It was not until 2008 that they asked:

> "If we make this loan, what is the likelihood that we will be repaid?"

EAST ASIA

The East Asian countries comprise the arc from Thailand to South Korea. According to K&A, the stimulus for an economic boom in the Asian countries in the 1990s was the implosion of the asset price bubble in Japan in 1990 and the appreciation of the yen that made investment in Asian countries more lucrative. As the Japanese stock and real estate markets imploded, funds that had been invested in Japan found their way into the East Asian countries.

In Thailand, Malaysia, and Indonesia stock prices increased by 300–500% in the first half of the 1990s and manufacturing activity surged. Real estate prices soared. The economies boomed.

K&A asserted that since the East Asian countries were quite dissimilar in many ways (e.g. Singapore, Taiwan, and Hong Kong were international creditor countries while Thailand and Malaysia were international debtors) there must have been some common factor causing the boom that overrode these differences.

As K&A pointed out, China, Thailand and the other East Asian countries profited from large-scale "outsourcing by American, Japanese, and European firms that wanted cheaper sources of supply for established domestic markets." America allowed its manufacturing capabilities to be transferred to these countries, with the payoff being importation of cheaper products made with lower cost labor in the East Asian countries. As these manufacturing facilities expanded in East Asia, they produced rapid economic growth, which in turn, led to more investment of foreign capital, particularly from Japan. Additional investment fed back into the booming economies producing ever more expansion.

The Japanese yen appreciated remarkably due to decades of trade surpluses, making it cheap for Japanese corporations to buy foreign currencies to buy or build subsidiaries in other Asian countries.[144] This allowed them to transfer production of standard manufactured products to subsidiaries abroad to take advantage of cheaper foreign labor. According to ZNET, the East Asian countries were the optimal target for outsourcing with their "disciplined work forces, low wages, pliable yet reliable governments," and the fact that there "was no need to worry about inadequate internal markets to buy the goods in the early years because the host governments agreed that the more goods destined for export the better."

> Neither Latin America, burdened by bad debt, nor stagnant African economies were attractive outlets for international capital. The former socialist economies in East Europe and the former Soviet Union were tempting, but not yet able to absorb large amounts of international capital quickly, and much riskier in any case. The East Asian tigers were simply the best investment opportunities in the late 1980s and early 1990s.[145]

As in many booms and bubbles, the initial basis for the East Asian boom was sound. Initially, the Asian export-oriented economies were competing with high-cost Western producers. Their "cheap workers, low taxes, and lax environmental laws" allowed them to under-price the competition and still earn good profits. But as more East Asian countries and businesses joined the export-led boom:

> The East Asian exporting economies competed more and more with each other rather than with Western producers, and . . . investments lost their luster and became less profitable than expected by both lender and borrower - a situation

that leads to problems in any highly leveraged credit system - even if there are no further complications.

To fuel this boom, East Asian countries needed to borrow foreign currencies to buy local currency and make loans to East Asian businesses at even higher interest rates. Asian banks earned high profits from a high volume of business conducted with a large spread between the interest rate they charged Asian businesses and the interest rate they paid international investors. As long as currency exchange rates remained stable, this could continue for some time:

> When competition among East Asian businesses led to falling export sales, these businesses could not repay their high interest loans from Asian banks. Moreover, falling export sales lowered international demand for the Asian currencies, leading to depreciation that made dollars more expensive for Asian banks to buy. For both reasons Asian banks could not repay their short-run dollar debts in the usual manner-by selling local currency from repaid loans for dollars.... When the Asian banks finally couldn't meet payments on their dollar loans it was too late. Their outstanding debt was too big and too short-term. As they scrambled to convert what local currency they had into dollars to meet their payment deadlines, they further depreciated the local currency. When the international investors and currency speculators and local wealthy elites caught on to what was happening, ... new dollar loans dried up overnight and more local currency was dumped on the exchange market, causing further depreciation.... At this point, there was no possibility of repaying international investors ... since the bottom had fallen out of the local currency making the dollars necessary for repayment prohibitively expensive. Moreover, factories couldn't produce exports for sale because they had no money to buy the imported inputs needed to make them, a condition made worse as the price of those inputs was multiplied due to depreciating local currencies.[146]

As K&A said:

> The nature of the bubble is that eventually it will be pricked, and then as with a child's balloon the air may escape sharply.

The bubbles in the Asian countries depended on a continual inflow of capital from foreign lenders. While the Asian economies were growing, currency exchange rates were stable and interest rates were attractive, so foreign money poured in. But as competition became more stringent, pressure built to devalue currencies as a means of making exports from the Asian countries more attractive. The devaluation of the Thai baht on July 2, 1997 was the first devaluation, and it led to what K&A called the "contagion effect."

> The depreciation of the baht triggered the contagion effect and within six months the foreign exchange values of each of the currencies on the Asian arc, with the exception of the Chinese yuan and the Hong Kong dollar, had lost 30 percent or more of their value in the foreign exchange markets. Stock prices declined by 30 to 60 percent, partly because foreign investors were seeking to cash out, partly because the domestic firms were no longer profitable. Real estate prices declined sharply. Most banks, with the exception of those in Singapore and Hong Kong, failed. The closing of many banks in Indonesia triggered racial strife, and an immense run on the currency that lost more than 70 percent of its value. When the crises occurred, the play script was a reprise of similar events in Japan in the previous decade. The chatter about the East Asian miracle disappeared.[147]

THE OIL BUBBLE

The world requires energy in order to operate. Prior to the 19th century, the prime source of energy was wood. By the mid-19th century, the forests of Europe were decimated, and it was fortunate that mankind discovered large deposits of coal and developed capabilities to replace wood with coal. But coal is not convenient for many of our needs, and it was only with the discovery of oil in the late 19th century, and the exploitation of oil deposits in the 20th century that the industrialization and modernizing of the developed nations could take place. The addition of natural gas in the 20th century provided additional momentum.

World production and consumption of oil expanded throughout the 20th century and on into the 21st century. In this process there was a widespread impression that oil resources were so huge and extensive that mankind could plunder them with impunity. However, one early voice warned of the consequences of depleting oil reserves as early as 1962. Like an Old Testament prophet waving his staff, M. King Hubbert (MKH) warned the world of the finiteness of petroleum reserves.[148]

There have been many attempts to estimate US and world oil reserves but most of these have considerable uncertainty because geological data are not as extensive and available as would be necessary. In addition, there is great political sensitivity to such activities.

Hubbert's fundamental postulate was that the exploitation of a resource proceeds through stages. In the early stages when production is low, there is a rapid increase in production as new markets are developed and new resources are discovered. Demand picks up and in the middle phase, production is able to roughly meet the demand. In the final stages, the resource is so depleted that production cannot keep up with demand. Data for the United States (including Alaska and offshore) are shown in Figure 2.16. Hubbert used the early data available in the 1960s and 1970s to estimate the ultimate production of US oil.[149] It is clear from the data that the US oil reserves are almost "fished out." With about 30 million barrels of oil remaining to be produced, and annual usage in the United States at roughly 7 billion barrels per year, there isn't much future in depending on US oil reserves. Interestingly, the Republicans in 2008 advocated additional drilling as the means to solve American energy problems. What could be more logical? If we have a shortage of oil, just drill for more! However, recent estimates for recoverable oil from Alaska North Slope are about 1 year's usage and from Florida offshore, less than one year's usage. (See inserts in Figure 2.16). Apparently, the Republicans never studied Hubbert.

Because of the depletion of US oil reserves, the United States has had to depend increasingly on imported oil. In 2008, the United States is consuming almost 7 billion barrels per year but is only producing about 2.5 billion barrels a year so it must import about 4.5 billion barrels per year. Unfortunately, much of the world's oil reserves are in the hands of nations that are ruled by despots, a number of which are antithetical to the United States.

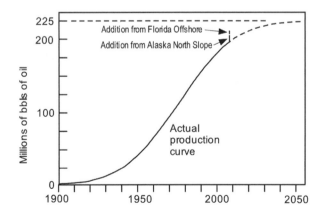

Figure 2.16 Cumulative production of US oil. (Presentation by Professor David Rutledge, Caltech, 2007).

The United States, with 5% of the world's population, accounts for about 25% of the world's oil consumption. The world's population is growing rapidly, going from about 2.5 billion in 1950 to about 6.6 billion in 2008. Most of that population growth has occurred in developing nations. A major goal of most of the projected 8.6 billion people in the developing nations (by 2050) is to live like Americans, which implies using a great deal more energy per capita than they use today. Coupling this projected future increase in energy per capita with future growth in population, leads to the conclusion that there will be significant increases in future world energy demand – barring a major world economic depression brought on by a lack of liquid fuels.

The situation for world oil resources is not as bad as it is for US oil resources. However, with present world oil consumption at around 27 billion barrels per year, and the expectation of future growth in demand, it appears likely that maximum production might occur sometime in the period 2010 to 2015. We can only guess about the future (see dashed estimates in Figure 2.17).

The historical data on world oil production are shown in Figure 2.18. Based on the cumulative curves shown in Figure 2.17, several projections can be made for future oil production as shown by dashed lines in Figure 2.18. The more that the world produces in the short run, the steeper will be the ensuing drop. Meanwhile, world demand for oil continues to rise. The US Energy Information Agency has projected world demand for oil to rise from 27 billion barrels per year in 2008 to 44 billion barrels per year in 2025.[152] Such a level of production would exceed the scale of Figure 2.18, and does not appear to be realistic. The expectation that the world can continue to

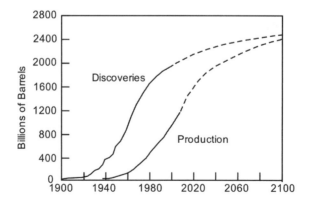

Figure 2.17 Cumulative world oil discoveries and production.[150]

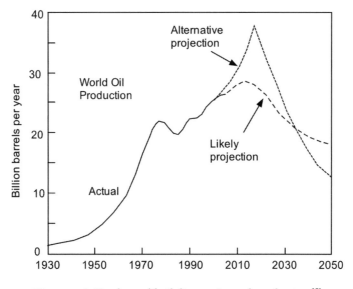

Figure 2.18 Yearly world oil discoveries and production.[151]

increase oil production forever has been a widespread fantasy, particularly in the United States.

Crude oil prices (unadjusted for inflation) are shown in Figure 2.19. Inflation from 1980 to 2007 was about a factor of almost 3, so that after adjustment for

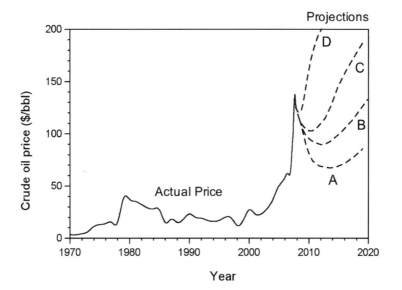

Figure 2.19 Crude oil price history.

inflation, the 2008 peak was only a little higher than the 1980 peak. A major question of interest is whether the steep rise in 2008 constitutes an artificial "bubble" created by speculators and market manipulators, or whether it represents the natural response to a major divergence between demand and supply. Many opinions have been voiced on both sides of the question. The steepness of the rise would tend to suggest an artificial bubble. In that case, one might expect curves "A" or "B" in Figure 2.19. However, the fact that demand is now beginning to push up against the natural limits of production, creates an environment provides a force driving prices up by virtue of supply and demand. While the price of oil has backed off in late 2008 from the highs set in early July 2008, due to a world economic deflation there will be continuing upward pressure on oil prices as developing nations industrialize. Thus, curves "C" and "D" are alternative possibilities for the future.

As of 2008, neither US presidential candidate, nor indeed the American public seems to appreciate the facts that the American oil resources are pretty much "fished out," the demand for liquid fuels is rising world-wide, and the situation can only get worse. As a result, there are proposals on the table for minor "band-aids" such as increasing domestic oil drilling, increasing inflation of tires[153] on vehicles, or lowering the speed limit on highways,[154] but no comprehensive long-term plan to provide the people with energy in the future. Governments tend to operate by setting quotas. For example, the State of California sets quotas for how much electric power will be produced by renewables by 2010, 2020, etc. There are calls in Congress and the U.N. for setting more quotas. These quotas are usually fantasies dreamed up by well meaning bureaucrats and idealists. But, how to get from here to there remains a mystery. The quotas are never met.

There is an energy crisis in the 21st century facing the world, and a liquid fuel crisis facing the United States. Government quotas will not solve anything. However, rising prices for energy may provide the impetus for innovators to face these problems. In that respect, higher oil prices may be the only catalyst to force us to face these problems.

Any solution to the world and US energy problems in the 21st century must take into account the fact that all fossil fuels (coal, oil, gas) produce carbon dioxide as the end product of combustion. Carbon dioxide is a greenhouse gas that contributes to the warming of the Earth. Since humans began burning fossil fuels in the 19th century, the concentration of carbon dioxide in the atmosphere has increased by 30%. It seems likely that if we continue to burn fossil fuels at projected rates, the carbon dioxide concentration by the end of the 21st century will probably be double that of the early 19th century. From 1900 to 2007, the average temperature of the Earth increased by about 0.8°C.[155] Many climatologists attribute most of this temperature rise to the increase in by CO_2 concentration during that period. Many are alarmed that further increases in carbon dioxide concentration will cause severe warming of the Earth, leading to severe impacts.[156] Assaulted by this barrage of claims of imminent disaster, well-meaning people have been misguided into believing that the fundamental challenge of the 21st century is to drastically reduce carbon emissions. As a result, proposals are being hatched to reduce future carbon dioxide emissions (on the assumption that carbon dioxide is responsible for global warming) but many of these proposals, if enacted, would produce a cataclysmic shortage of fuels and a major world economic depression.

Actually, the degree to which rising carbon dioxide concentration has contributed to rising Earth temperature is not well established.[157] As the carbon dioxide concentration builds up, each addition produces a smaller rise in temperature. The climate models depend on many assumptions that have not been validated. Nevertheless, the prevailing view in many government science agencies is that global warming is a bigger problem than the challenge of providing people with energy – a major misconception. In October 2007, Chairman Dingell's Energy and Commerce Committee of the US Congress released its first white paper in a series on "Meeting the Climate Change Challenge." This paper advocated:

> The United States should reduce its greenhouse gas emissions by between 60 and 80 percent by AD 2050 to contribute to global efforts to address climate change.

If the reduction was meant to be from present levels, the reduction from business as usual in 2050 would be even greater: 72 to 86%. This proposal does not require that the United States supply its people and its industries with the energy needed to operate. It merely requires that emissions be reduced. An epidemic of insanity has apparently invaded the US House of Representatives. Such a program will send the United States reeling back toward the lifestyle of the 18th century, bringing on a far worse economic depression than that of 1930s.

The fundamental challenge of the 21st century is to provide the burgeoning population of the world with energy (also food, water, . . .) without desecrating the environment. To the extent that the world can provide these resources and at the same time reduce carbon emissions, that is a good thing. However, expenditure of large amounts of energy to sequester CO_2 will only exacerbate the energy problem.

The oil bubble of 2008 has provided a preview of the coming energy crisis in the 21st century. So far, we have not fully appreciated the extent of the challenge that lies ahead.

THE NEXT BUBBLE

Eric Janszen provided valuable insights into bubble formation and popping.[158] Janszen's view is that major industries like steel and autos no longer dominate the economy. According to him, "the new economy belongs to finance, insurance, and real estate—FIRE." He described FIRE as "a credit-financed, asset-price-inflation machine" that is built upon a fundamental belief that the value of one's assets no longer fluctuates in response to the business cycle and the financial markets, but now mainly rises, with only infrequent short-term reversals.

Janszen provided an answer to a question: Why do foreigners invest in US securities when we borrow rampantly and owe so much debt? As Janszen explained, the United States has a severe trade imbalance with oil-producing countries, Japan and more recently, China. The question is what should these countries do with the dollars that keep piling up in their coffers? The United States provides military protection to countries like Saudi Arabia and Japan. In addition, China and other countries need to support the United States, because the United States provides a critical market for their goods and provides world stability. So, for a variety of reasons, most countries with favorable trade balances with the United States are motivated to continually invest acquired dollars in US assets. If they did not, the value of the dollar would fall precipitously, and that would reduce the value of their dollar holdings, and reduce the ability of the United States to import their products. Janszen quotes an old proverb that says if you owe a bank a small amount, the bank controls you; but if you owe the bank more than it can afford to lose, then you control the bank. He says that the United States owes so much to foreign countries that these countries must continually prop up US assets. However, the US policies of cutting taxes, raising expenditures, using gas-eating cars, spending trillions on Iraq, handing out money to its citizens (that it does not have), and generally recklessly borrowing, has put these foreign investors to a severe test.

Janszen provided some intriguing graphical depictions of bubbles. I have taken the liberty of modifying his graphs. Figure 2.20 shows a revised version of Janszen's graph for the total market value of NASDAQ stocks. However, one caveat that should be borne in mind is that during the heyday of the dot.com boom (late 1990s to 2000) the great preponderance of NASDAQ

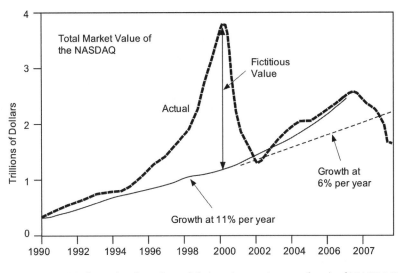

Figure 2.20 Market value (number of shares times price per share) of NASDAQ stocks vs. year.

stocks were closely held and only a small fraction of outstanding stocks were actively traded on the markets. Therefore, multiplying the number of shares by the current price (as was done to obtain Figure 2.20) is misleading. There is no way that the price could have been maintained if most of the shares were put on the market. The shares were maintained artificially high because of the small amount available for purchase by the mob. Thus, the peak shown in Figure 2.20 is labeled "fictitious."

Janszen compared the actual NASDAQ history with a curve representing 11% growth per year. However, Janszen did not include the fall-off in the NASDAQ in 2008 that seems to suggest a long-term growth rate of perhaps 6%. It is likely that the termination of each decline following a bubble will overshoot on the downward side.

Similarly, I have modified Janszen's curve for the market value of US real estate, as shown in Figure 2.21. If his projection for the future (dashed line) proves to be accurate, real estate has a much deeper drop in store than any market analyst has predicted.

Finally, Janszen projected forward into the future. Figure 2.22 shows the projected increase in asset values above the long-term trend. The $3 trillion NASDAQ bubble peaked in 2000, and the $12 trillion real estate bubble peaked in 2006–2007. Hence projecting ahead, Janszen suggested that the next bubble is likely to be focused on alternative energy sources, and it may be expected to

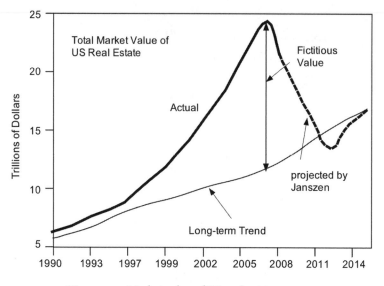

Figure 2.21 Market value of US real estate vs. year.

peak in 2013–2014 with a fictitious valuation of perhaps up to $20 trillion. Obviously, this is a shot in the dark. If anyone could actually predict the nature, size and timing of the next bubble, it would be a sure pathway to riches. That there will be bubbles in the future seems likely. The "something for nothing

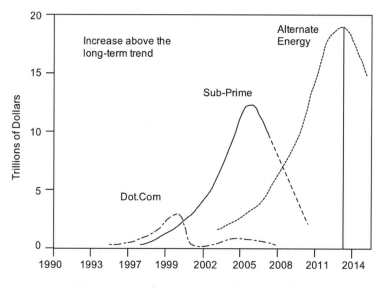

Figure 2.22 Is alternate energy the next bubble?

generation" will bid some commodity or market up to incredible heights. But in the meantime, the United States (and the world as well) has to deal with basic issues:

- The US rate of spending and the tax structure are out of balance. This requires that the United States will need to continue to borrow heavily merely to provide annual operating expenses. There does not seem to be any credible way to reduce spending adequately, although an early pullout from Iraq would help. Nor is there any credible way to increase revenues except by significantly increasing taxes, but that appears to be politically unacceptable. Continued borrowing for operating expenses with interest payments made from new borrowing constitutes a classical Ponzi scheme. The only reasons that the world puts up with this situation are that they need our markets and our military protection.

- The prosperity of the past several years has been based primarily on paper asset growth, rather than real wage growth. Now that the real estate market is headed downward precipitously, and the stock markets have dropped and vacillated, this economic stimulus via the "wealth effect" appears to be rapidly diminishing. Consumer sentiment is at the lowest it has been in decades. It is likely that consumer spending will tank in the next year or two, bringing on a significant recession.

- By historical standards, stocks remain very highly priced in 2008, and are vulnerable to further downward trends.

- The annual US production of petroleum has been decreasing for years and will continue to decrease, whether the Alaskan North Slope or Florida Offshore regions are exploited or not. The price of fossil fuels will increase, providing a stifling influence on the economy. Meanwhile, our imports continue to grow and our preparation for future energy production is myopic.

- The cost of social security will increase with the aging population, and this will require higher social security taxes.

- The levels of debt in the United States at the federal, state, municipal and individual level are excessive, and will have to be dealt with sooner or later.

Therefore, there may be a long deep valley that must be traversed before the next bubble rises to euphoric heights.

NOTES

[1] Stock Market Crashes, Productivity Boom Busts and Recessions: Some Historical Evidence, Michael Bordo, January, 2003.

[2] G. Wood, "Great Crashes In History: Have They Lessons For Today?" Oxf Rev Econ Policy 1999; 15:98–109 http://ideas.repec.org/a/oup/oxford/v15y1999i3p98-109.html.

[3] World Economic Outlook, IMF, April, 2003 http://www.imf.org/external/pubs/ft/weo/2003/01/pdf/chapter2.pdf

[4] No disrespect meant for George C. Scott.

[5] B. Mark Smith, "A History of the Global Stock Market," University of Chicago press, 2003.

[6] B. Mark Smith, "A History of the Global Stock Market," University of Chicago press, 2003.

[7] Fredrick Lewis Allen, "Only Yesterday", Harper and Row, 1931.

[8] Fredrick Lewis Allen, *loc cit.*

[9] Note that during the sub-prime real estate boom of 2002–2007, many properties were sold for no money down, and some were sold on the basis that no payments of principal need be made for the first few years. Thus the 1920s Florida land boom was in some ways conservative compared to the 21st century real estate boom.

[10] The Economic Boom of the 1920s, Alex Aylen, http://www.planetpapers.com/Assets/3950.php

[11] Post War Boom and Bust (1920–1929), http://www.flowofhistory.com/category/export/html/151

[12] I attended a lecture by a noted advocate of free trade, who explained how one country would produce autos, and another would produce TV sets, etc. Someone in the audience raised the question: "What will the United States do?" Another person in the audience shouted: "Consume!"

[13] John Kenneth Galbraith, "The Great Crash" Mariner Books, 1954.

[14] I am reminded of a New Yorker cartoon that shows the manager of a professional baseball team going out to the mound to talk to his pitcher. In the background, one can perceive a stock market ticker in the dugout, and the conversation at the mound is not about baseball, but about investments.

[15] "What is the Link Between Margin Loans and Stock Market Bubbles?" Markus Ricke, http://papers.ssrn.com/sol3/papers.cfm?abstract_id=473781

[16] "Monetary Policy Should Gently Lean against Bubbles" in "Irrational Exuberance", Robert J. Shiller, Doubleday, 2nd ed., 2004.

[17] In late 1929, margin interest rates often approached 20% and investors were happy to pay that.

[18] Robert J. Samuelson, "The Great Depression", http://www.econlib.org/Library/Enc/GreatDepression.html

[19] Christine Romer, Encyclopedia Brittanica 2003, http://elsa.berkeley.edu/~cromer/great_depression.pdf

[20] Stock Market Crashes, Productivity Boom Busts and Recessions: Some Historical Evidence, Michael Bordo, January, 2003.

[21] J. K. Galbraith, "The Affluent Society", *loc cit.*

[22] Milton Friedman and Anna Schwartz "A Monetary History of the United States, 1867–1960", Princeton University Press, 1971.

[23] "The Savings and Loan Crisis and Its Relationship to Banking". FDIC Report, 1996. http://www.fdic.gov/bank/historical/history/167_188.pdf

[24] "High Rollers - Inside the S&L debacle", Martin Lowy, Praeger, 1991; "In$side Job - the looting of America's S&Ls", Stephen Pizzo, Mary Fricker and Paul Muolo, Harper Perennial, 1989.

[25] FDIC Report, *loc cit.*

[26] At the height of the OPEC oil embargo in 1973, when the economy was in dire straits, Congress passed two pieces of legislation to combat the shortage of oil: (1) the 55 mph speed limit, and (2) extended daylight savings time. Neither policy had any significant effect. But Congress had shown that it could take "action."

[27] This also happened in 2007–2008 when Countrywide Financial was wracked by losses from sub-prime loans, and in an effort to raise capital, it offered higher CD rates than any other bank. In fact, with shaky banks offering higher interest rates on deposits with FDIC backing, it is advantageous to put one's savings into shaky banks.

[28] FDIC Report, *loc cit.*

[29] The ironic thing is that the politicians behind this legislation were Republican "states rights" advocates.

[30] The stupidity of the Reagan administration was gargantuan, and compounded by twisting their view of reality to fit their internal philosophies; yet Mr. Reagan is widely regarded as a hero, and the Republican candidates for president in 2008 vied with one another in asserting they were the most like Mr. Reagan.

[31] FDIC Report, *loc cit.*

[32] "High Rollers - Inside the S&L debacle", Martin Lowy, Praeger, 1991

[33] Martin Lowy, *loc cit.*

[34] It is noteworthy that in such a case, the actual payments by the developer prior to completion of the project were nil. Nevertheless, the S&L claimed the loan origination fee as "income" which it paid to itself out of depositors' money. This was clearly a form of Ponzi scheme.

[35] FDIC Report, *loc cit.*

[36] Michael Lewis, "Liar's Poker," Penguin Books, 1989.

[37] "In$side Job - the looting of America's S&Ls", Stephen Pizzo, Mary Fricker and Paul Muolo, Harper Perennial, 1989.

[38] "In$side Job - the looting of America's S&Ls", Stephen Pizzo, Mary Fricker and Paul Muolo, Harper Perennial, 1989.

[39] "In$side Job - the looting of America's S&Ls", *loc cit.*

[40] Lincoln Savings And Loan, Mary Nisbet, and Donald R. Loster, AICPA Case Development Program Case No. 96–05, http://www.aicpa.org/download/edu/96-05a.pdf

[41] Stephen Pizzo, Mary Fricker and Paul Muolo, *loc cit.*

[42] Stephen Pizzo, Mary Fricker and Paul Muolo, *loc cit.*

[43] "The Keating Five", Dan Nowicki, Bill Muller, The Arizona Republic, Mar. 1, 2007.

[44] I am reminded of a New Yorker cartoon showing a parent changing a flat tire in the rain, explaining to the children in the car that he could not change to another channel because this was really happening.

[45] Maggie Mahar, "Bull" - A History of the Boom, 1982–1999, Harper Business, 2003.

[46] See "How Mr. Reagan Made a Bad Problem Worse, The Crash of 1987 and Enron", in Chap. 2 and Figures 2.3 and 2.4 in "Income Tax Brackets and Budget Deficits" in Chap. 1.

[47] Martin Lowy, "High Rollers - Inside the S&L debacle," Praeger, 1991.

[48] Greenspan and the Federal Reserve could not discern a bubble.

[49] Greenspan enacted 24 consecutive rate cuts from 1989 to 1992.

[50] Actually, prior to the great bull market of 1982–1999, stocks were not such a good investment, and only the 1982–1999 bull market made stocks look good in retrospect, looking backward from after 1999.

[51] According to Maggie Mahar, "mergers, takeovers and leveraged buyouts from 1984 through 1987 slashed the supply of stock available on the open market by > $250 billion. By 1988 121 firms in the S&P 500 had vanished."

[52] The tax code provided a very generous treatment of interest payments on debt. Interest payments on debt were fully deductible. Buying assets with borrowed funds meant shifting much of the cost to the federal government.

[53] B. Mark Smith, A History of the Global Stock Market, University of Chicago press, 2003.

[54] This echoed the support from economists that was prevalent in the late 1920s.

[55] Indeed, they may be right. If prosperity can result from heavy borrowing and bidding up the price of paper assets, as it has, maybe the old rules of economics no longer apply after all.

[56] John Cassidy, "Dot.Con" Perennial Press, 2002.

[57] John Cassidy, "Dot.Con" Perennial Press, 2002.

[58] John Cassidy, *loc cit.*

[59] The BLS website provides extensive data on productivity by sector and industry. The picture is highly complex and beyond the scope of the present book. http://www.bls.gov/lpc/#data

[60] "A Brief History of the 1987 Stock Market Crash with a Discussion of the Federal Reserve Response", Mark Carlson, Staff Working Paper 2007–2013, http://www.federalreserve.gov/pubs/FEDS/2007/200713/200713pap.pdf

[61] "Endogenous Risk", Jon Danielsson and Hyun Song Shin, April 21, 2002, http://hyunsongshin.org/www/risk1.pdf

[62] "The Big, Bad Wolf and the Rational Market: Portfolio Insurance, the 1987 Crash and the Performativity of Economics", Donald MacKenzie, February 2004 http://www.sociology.ed.ac.uk/Research/Staff/Mackpaper5.pdf

[63] "Stock Market Crash: A History of Financial Train Wrecks", http://www.stock-market-crash.net/nasdaq.htm

[64] "Dot.Con", John Cassidy, Harper Perennial Books, 2002.

[65] "Stock Market Crash: A History of Financial Train Wrecks", *loc cit.*

[66] Maggie Mahar, "Bull! - A History of the Boom," 1982–1999, Harper Business, 2003, page xx.

[67] Maggie Mahar, *loc cit.*

[68] As we pointed out earlier, President Truman wanted a one-handed economist.

[69] Actually, this is one of the few places where Greenspan's ideas made sense to me.

[70] In a speech on January 13, 2000 in New York.

[71] This, in itself is amazing and seems to defy Economics 101. With all the excess money generated by huge profits in the stock market, why wasn't there a significant inflation? It seems as if wealth could be created out of nothing by simply bidding up the price of paper (actually, stock certificates) and there was no penalty to be paid in rising inflation. Perhaps the answer to this conundrum is that during this period, America was busily transferring almost all its manufacturing capabilities to China and other Asiatic countries, so they could send us cheap goods and thus keep a lid on inflation. Such a scenario would not last forever, but it seemed to apply during the 1998–2000 period.

[72] Senator Gramm was not noted for his intellect or clarity of thought. But he did distinguish himself by getting rich through support of the S&L bubble (see "The Savings and Loan Scandal of the 1980s" in this chapter).

[73] Maggie Mahar, "Bull" - A History of the Boom, 1982–1999, Harper Business, 2003, page xix.

[74] Maggie Mahar shows that Blodgett had severe doubts about the staying power of the dot.com craze but his career depended on not voicing these views. In fact, Mr. Blodgett received death threats for every hint that the dot.com stocks were overpriced. *loc cit.*, p.xxi.

[75] "Unconventional Success: A Fundamental Approach to Personal Investment" by David Swensen, Free Press, 2005; thanks to Giulio Varsi for bringing this to my attention.

[76] United States District Court, Southern District of New York, Master File No. 02 MDL 1484, Judge Milton Pollack.

[77] http://securities.stanford.edu/1024/MER02-01/index.html

[78] "Adelphia Comes Clean", CFO Magazine, http://www.cfo.com/article.cfm/3011051/c_3046603?f=insidecfo

[79] Barings Debacle, Risk Glossary, http://www.riskglossary.com/link/barings_debacle.htm

[80] Sungard Bancware eRisk, http://www.erisk.com/Learning/CaseStudies/AlliedIrishBanks.asp

[81] http://wcbstv.com/national/Societe.Generale.fraud.2.638859.html

[82] "Price-gouging Inquiries Target Enron Overcharges in California May Exceed $40 Billion," By Kathleen Sharp, Boston Globe Correspondent, Mar 03, 2002.

[83] Wikipedia

[84] Kathleen Sharp, *loc cit.*

[85] Kathleen Sharp, *loc cit.*

[86] The Wikipedia quotes Kenneth Lay as saying: "In the final analysis, it doesn't matter what you crazy people in California do, because I got smart guys who can always figure out how to make money."

[87] "Why Enron Went Bust", Fortune, 12/21/01, Bethany McLean.

[88] "Blind Faith: How Deregulation and Enron's Influence Over Government Looted Billions from Americans" – Sen. Gramm, White House Must Be Investigated for Role in Enron's Fraud of Consumers and Shareholders, December 2001, Public Citizen's Critical Mass Energy & Environment Program, http://www.citizen.org.

[89] Kenneth Lay was a visitor to the White House where he was a pal of George W. Bush.

[90] It is noteworthy that Mr. Gramm surfaced again in July 2008 with his commentary on the financial crisis declaring that we have become "a nation of whiners." The Republican Party hastened to disassociate itself from Mr. Gramm. ("McCain doesn't need enemies. He has friends," Newsweek, July 10, 2008.)

[91] When Genius Failed: The Rise and Fall of Long Term Capital Management, Roger Lowenstein, Random House, 2000.

[92] Unless this clarification shows the investment to be worse than was initially supposed.

[93] "Endogenous Risk", Jon Danielsson and Hyun Song Shin, September 21, 2002, hyunsongshin.org/www/risk1.pdf

[94] "LTCM Speaks", Joe Kolman, http://www.derivativesstrategy.com/magazine/archive/1999/0499fea1.asp

[95] Nassim Taleb, "The Black Swan: The Impact of the Highly Improbable", Random House, 2007.

[96] "Endogenous Risk", Jon Danielsson and Hyun Song Shin, September 21, 2002, http://hyunsongshin.org/www/risk1.pdf

[97] "LTCM Speaks", Joe Kolman, http://www.derivativesstrategy.com/magazine/archive/1999/0499fea1.asp

[98] "Too Big to Fail? Long-Term Capital Management and the Federal Reserve", Kevin Dowd, Cato Briefing Paper 52, September 23, 1999.

[99] Kevin Dowd, *loc cit.*

[100] Kevin Dowd, *loc cit.*

[101] "Post-Socialist Financial Fragility: the Case of Albania", Dirk J. Bezemer, http://www.tinbergen.nl/discussionpapers/99045.pdf

[102] Dirk J. Bezemer, *loc cit.*

[103] PricewaterhouseCoopers, Deloitte Touche Tohmatsu, KPMG International, Ernst & Young, and Arthur Andersen, each with revenues in 2001 that exceeded $10 billion.

[104] "Lack of Accountability: The Enron/Arthur Andersen Scandal and the Future of The Accounting Business", Philip Mattera, Corporate Research E-Letter No. 21, February 2002. http://www.corp-research.org/archives/feb02.htm

[105] Phillip Mattera, *loc cit.*

[106] Phillip Mattera, *loc cit.*

[107] It seems probable that the large accounting firms were willing to pay fines and damages that were far less than their profits. This seems to be analogous to automobile companies that would rather pay damages from suits, than pay the cost of making their cars safe.

[108] Skeptical CPA, http://skepticaltexascpa.blogspot.com/

[109] Supreme Court Inc., New York Times Magazine, March 16, 2008.

[110] James B. Stewart required almost 600 pages to tell the whole story in his book: "Den of Thieves" Simon and Schuster, 1991.

[111] There is a saying on Wall Street that "The bulls make money; the bears make money; and the pigs get eaten."

[112] In this context, "parking violations" has nothing to do with vehicles. It refers to using surreptitious ownership of a stock through an intermediary to hide the true ownership.

[113] With this huge overhang of low-rate long-term mortgages, the American banking industry is susceptible to future increases in interest rates that could drive down the spread between the rates they collect on mortgages and the rates they pay out on deposits. A significant future rise in interest rates could wipe out the whole banking industry. Perhaps that is why the Federal Reserve works so hard to keep a lid on interest rates.

[114] Note: The 10–20% profit is on the house price. For an investor who puts 5% down on the house, the profit *on his investment* is 200–400%. For an investor who puts no money down, the profit margin is infinite.

[115] Wall Street Journal (February 6, 2008) http://online.wsj.com/article/SB120225852 189145889.html?mod=todays_us_marketplace

[116] http://www.CNNMoney.com (June 5, 2008)

[117] California Association of Realtors; http://www.doctorhousingbubble.com

[118] World Economic Outlook, IMF, April, 2003 http://www.imf.org/external/pubs/ft/weo/2003/01/pdf/chapter2.pdf

[119] Finance and Economics Discussion Series, Divisions of Research & Statistics and Monetary Affairs, Federal Reserve Board, Washington, D.C., Sources and Uses of Equity Extracted from Homes, Alan Greenspan and James Kennedy, 2007–2020.

[120] http://www.futurefocus.net/news/newsoct05.htm

[121] http://www.federalreserve.gov/boarddocs/speeches/2003/20030304/default.htm

[122] The Effects of Recent Mortgage Refinancing, Peter J. Brady, Glenn B. Canner, and Dean M. Maki, Federal Reserve Bulletin July 2000; Mortgage Refinancing in 2001 and Early 2002, Glenn Canner, Karen Dynan, and Wayne Passmore, of the Board's Division of Research and Statistics, Federal Reserve Bulletin December 2002.

[123] Charles Steindel, "How Worrisome Is a Negative Saving Rate?" Federal Reserve Bank of New York, Current Issues, May 2007.

[124] The Market Oracle, http://www.marketoracle.co.uk/Article307.html

[125] William Poole, President of the Federal Reserve Bank of St. Louis in a speech entitled: "Real Estate in the U.S. Economy" before the Industrial Asset Management Council Convention in St. Louis on Oct. 9, 2007.

[126] Dr. Housing Bubble, http://www.doctorhousingbubble.com

[127] Dr. Housing Bubble, *loc cit.*

[128] Med Yones, "U.S. Economy Risks and Strategies for 2007–2017", http://iim-edu.org/u.s.economyrisks/

[129] http://online.wsj.com/article_print/SB120243369715152501.html

[130] Los Angeles Times, February 14, 2008.

[131] Los Angeles Times, July 23, 2008.

[132] Bloomberg.com

[133] Bloomberg.com

[134] e.g. http://www.creditwritedowns.com/2008/05/credit-crisis-timeline.html

[135] "Unbridled Markets: Conservatives Embrace Securitization Run Amok", Scott Lilly, http://www.americanprogress.org/issues/2007/12/unbridled_markets.html

[136] Scott Lilly, *loc cit.*

[137] Lilly put the word "homeowners" in quotes, possibly because their equity was so small that they could hardly be called owners, or possibly because many of them were speculators and did not live in the home.

[138] "Triple-A Failure," Roger Lowenstein, *loc cit.*

[139] http://michellemalkin.com/2007/12/03/hillary-and-bush-agree-government-should-bail-out-homeowners/

[140] http://www.foreignpolicy.com/story/cms.php?story_id=3976 (reporting on Morgan-Stanley data).

[141] Floyd Norris, New York Times, April 5, 2008, "OFF THE CHARTS: Across the Globe, Hints of More Perils in Housing."

[142] K&A, *loc cit.*

[143] B. Mark Smith, "A History of the Global Stock Market," University of Chicago press, 2003.

[144] "What Actually Turned the Asian Boom into Bust?" ZNET, http://www.zmag.org/Instructionals/GlobalEcon/id13_cf.htm

[145] ZNET, *loc cit.*

[146] ZNET, *loc cit.*

[147] K&A, *loc cit.*

[148] K. S. Deffeyes, *Hubbert's Peak* - The Impending World Oil Shortage, Princeton University Press, 2001.

[149] D. Rapp, "Estimation of degree of advancement of petroleum exploration in the US," Energy Sources 2, 125–146 (1975).

[150] Seppo A. Korpela, "Oil depletion in the world," Current Science 91, 1148–1152 (2006); http://www.greatchange.org/ov-korpela,US_and_world_depletion.html

[152] DOE EIA, International Energy Outlook, 2004. Robert L. Hirsch, Roger Bezdek, Robert Wendling, "Peaking Of World Oil Production: Impacts, Mitigation, & Risk Management," February 2005 (www.netl.doe.gov/publications/others/pdf/Oil_Peaking_NETL.pdf)

[151] S. Korpela, *loc cit.*

[153] Obama pumped up over attack on his energy strategy http://www.latimes.com/news/nationworld/washingtondc/la-na-campaign6-2008aug06,0,2054832.story

[154] http://speier.house.gov/apps/list/press/ca12_speier/60mph.shtml. Note that in California, the typical vehicle speed on freeways is 80+ mph whereas the speed limit is 65 mph. Is there any reason to believe that dropping the speed limit to 60 mph would change anything?

[155] Donald Rapp, *Assessing Climate Change*, Praxis Publishing, 2008.

[156] The UN's Intergovernment Panel of Climate Change took over 1,000 pages to describe all the putative horrendous impacts of future global warming.

[157] D. Rapp, *Assessing Climate Change, loc cit.*

[158] Eric Janszen, "The Next Bubble: Priming the markets for tomorrow's big crash," Harper's Magazine, February, 2008.

Abbreviations

ACC	American Continental Corporation
ADC	Acquisition, development, and construction
AMT	Alternative minimum tax
AOL	America On-Line
ARM	Adjustable rate mortgage
ARPA	Advanced Research Projects Agency
ATM	Automatic teller machine
BAPCPA	Bankruptcy Abuse Prevention and Consumer Protection Act of 2005
BLS	Bureau of Labor Statistics
CBPP	Center for Budget and Policy Priorities
CD	Certificate of Deposit
CEO	Chief executive officer
CFO	Chief financial officer
CFTC	Commodity Futures Trading Commission
CINB	Continental Illinois National Bank and Trust Company
CPI	Consumer price index
CPI-U	Consumer price index for urban areas
CSREI	Case-Shiller Real Estate Index
DBL	Drexel, Burnham and Lambert
DCJ	David Cay Johnston
DIDC	Depository Institutions Deregulation and Monetary Control Act of 1980
DJIA	Dow-Jones industrial average
DOJ	Department of Justice
ENW	Edward N. Wolff
ERISA	Employee Retirement Income Security Act

FDIC	Federal Deposit Insurance Company
Fed	Federal Reserve System
FERC	Federal Energy Regulatory Commission
FHLBB	Federal Home Loan Bank Board
FOMC	Federal Open Market Committee
FSLIC	Federal Savings and Loan Insurance Corporation
GDP	Gross Domestic Product
GNP	Gross National Product
GPS	Global positioning system
IMF	International Monetary Fund
IPO	Initial public offering
JKG	John Kenneth Galbraith
K&A	Kindleberger and Aliber
LTCM	Long Term Capital Management
LTV	Loan-to-value ratio
MBLI	Mutual Benefit Life Insurance
NBR	Nightly Business Report
MWH	Megawatt-hours
NBR	Nightly Business Report (Public Television)
NYSE	New York Stock Exchange
NYTI	New York Times Index (of 25 industrial stocks)
OMB	Office of Management and Budget
OPEC	Organization of Petroleum Exporting Countries
P/E	Price/earnings ratio
PBGC	Pension Benefit Guaranty Corporation
PCE	Personal consumption expenditures
PFM	Pizzo, Fricker and Muolo
PG&E	Pacific Gas and Electric
PPI	Producer price index
S&L	Savings and loan
S&P	Standard and Poor's
SEC	Securities and Exchange Commission
SIV	Structured investment vehicle
SS	Social Security
TIAA-CREF	Teachers Insurance Annuity Association
WTHII	Inflation-what-the-heck-is-it
WWII	World War II
Y2K	Year 2000 problem

Index

Index

Index